Field Manual
NO. 3-23.35

HEADQUARTERS
DEPARTMENT OF THE ARMY
WASHINGTON, DC, 18 April 2007

COMBAT TRAINING 9mm PISTOLS

CONTENTS

Distribution Restriction: Approved for public release. Distribution is unlimited.

*This publication supersedes FM 23-35, 10 October 1988.

PREFACE

*This publication applies to the Active Army, the Army National Guard (ARNG)/Army National Guard of the United States (ARNGUS), and the US Army Reserve (USAR) unless otherwise stated. It provides guidance on the operation and marksmanship of the M9, 9-mm pistol and the M11, 9-mm pistol. It reflects current Army standards in weapons qualification. It is a guide for the instructor to develop training programs, plans, and lessons that meet the objectives of the US Army Marksmanship program for developing combat-effective marksmen. The Soldier develops confidence, knowledge, and skills by following the guidelines in this manual.

Unless this publication states otherwise, masculine nouns and pronouns refer to either gender. Also, some of the uniforms illustrated in this manual are shown without camouflage for greater clarity.

*This publication prescribes DA Form 88-R, *Combat Pistol Qualification Course Scorecard,* and DA Form 5704-R, *Alternate Pistol Qualification Course Scorecard.*

*The proponent for this publication is the US Army Training and Doctrine Command (TRADOC). The preparing agency is the US Army Infantry School. You may send comments and recommendations by any means, US mail, e-mail, fax, or telephone, as long as you use or follow the format of DA Form 2028, *Recommended Changes to Publications and Blank Forms*. You may also phone for more information.

E-mail	229-DOC-LIT@conus.army.mil
Phone	COM (706) 545-8623 or DSN 835-8623
Fax	COM (706) 545-8600 or DSN 835-8600
US Mail	Cmdt, USAIS ATTN: ATSH-INB 6650 Wilkin Drive, Bldg 74, Rm 102 Fort Benning, GA 31905-5593

CHAPTER 1 *

COMPONENTS AND FUNCTIONING

This chapter describes the M9 and M11 semiautomatic pistols, their maintenance requirements, and their operation and functioning.

Section I. DESCRIPTION AND COMPONENTS

The M9 (Figure 1-1) and M11 (Figure 1-2) pistols are 9-mm, semiautomatic, magazine-fed, recoil-operated, double-action weapons. Both are chambered for the 9-mm cartridge.

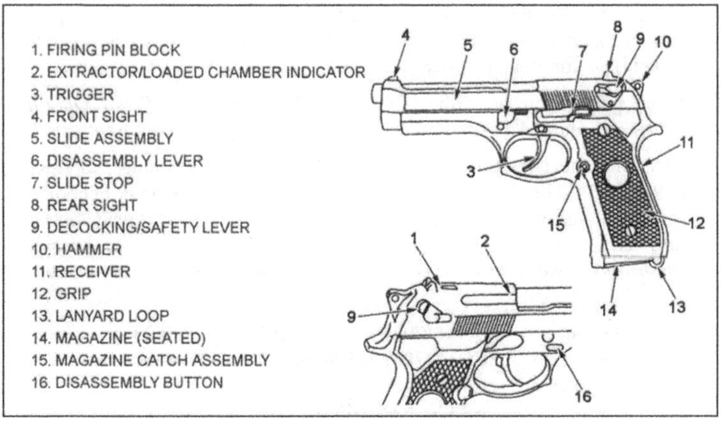

Figure 1-1. 9-mm pistol, M9.

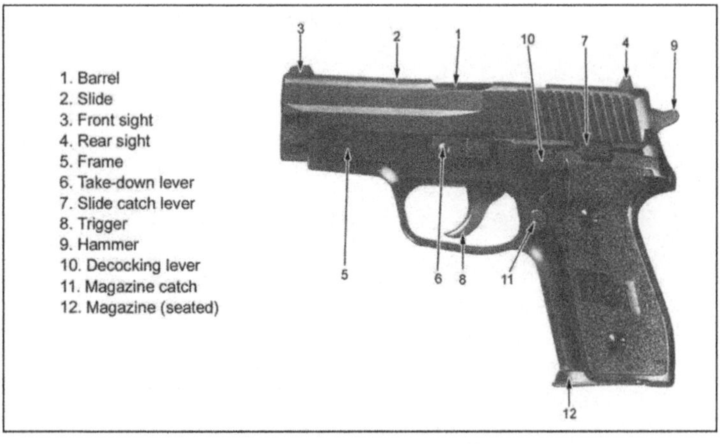

Figure 1-2. 9-mm pistol, M11.

1-1. DESCRIPTION

Table 1-1 summarizes data for both pistols.

	M9 PISTOL	M11 PISTOL
Caliber	9-mm NATO	9-mm NATO
System of Operation	Short recoil, semiautomatic	Short recoil, semiautomatic
Locking System	Oscillating block	Oscillating block
Length	217 mm (8.54 inches)	180 mm (7.08 inches)
Width	38 mm (1.5 inches)	37 mm (1.46 inches)
Height	140 mm (5.51 inches)	136 mm (5.35 inches)
Magazine Capacity	15 Rounds	13 Rounds
Weight with Empty Magazine	960 g (2.1 lb)	745 g (1.6 lb)
Weight with Magazine Fully Loaded	1,145 g (2.6 lb)	830 g (1.8 lb)
Barrel Length	125 mm (4.92 in)	98 mm (3.86 in)
Rifling	Right-hand, six-groove (pitch 250 mm [about 10 in])	Right-hand, six-groove (pitch 250 mm [9.84 in])
Muzzle Velocity	375 meters/second (1,230.3 feet/second)	375 meters/second (1,230.3 feet/second)
Muzzle Energy	569.5 Newton meters (420 foot pounds)	569.5 Newton meters (420 foot pounds)
Maximum Range	1,800 meters (1,962.2 yards)	1,800 meters (1,962.2 yards)
Maximum Effective Range	50 meters (54.7 yards)	50 meters (54.7 yards)
Front Sight	Blade, integral with slide	Blade, dovetailed to slide
Rear Sight	Notched bar, dovetailed to slide	Notched bar, dovetailed to slide
Sighting Radius	158 mm (6.22 inches)	145 mm (5.8 inches)
Safety Features	Decocking/safety lever, firing pin block.	Decocking lever, automatic firing pin lock.
Hammer M9 (half-cocked notch) Hammer M11 (cocked or decocked)	Prevents accidental discharge. Stops the hammer from accidentally falling fully forward if the full-cock notch is stripped. Also, catches the hammer if the cocking cycle is interrupted in the double-action fire mode.	Prevents accidental discharge. Stops the hammer from accidentally falling fully forward if the full-cock notch is stripped. Also, catches the hammer if the cocking cycle is interrupted in the double-action fire mode.
Basic Load, Three Magazines	45 rounds	39 rounds
Trigger Pull	Single-action: 5.50 pounds Double-action: 12.33 pounds	Single-Action: 4.40 pounds Double-Action: 12.12 pounds

Table 1-1. Data for the M9 and M11 pistols.

NOTE: For additional information on technical aspects of the two pistols, see the following TMs:

M9: TM 9-1005-317-10
M11: TM 9-1005-325-10

WARNING

If you release the hammer while manually cocking the pistol (but *do not touch* the trigger), then the half-cocked position will catch the hammer and prevent the pistol from firing.

However, *never* use the half-cocked position as a safety. Also, if you pull the trigger with the pistol half-cocked--

THE PISTOL *WILL* FIRE.

1-2. COMPONENTS

The major components of the M9 (Figure 1-3) and M11 (Figure 1-4, page 1-4) pistols include--

a. **Slide and Barrel Assembly**: Houses the firing pin, striker, and extractor. Cocks the hammer during recoil cycle.

b. **Recoil Spring and Recoil Spring Guide**: Absorbs recoil and returns the slide assembly to its forward position.

c. **Barrel and Locking Block Assembly M9/Barrel Assembly M11:** Houses cartridge for firing, directs projectile, and locks barrel in position during firing.

d. **Receiver**: Supports all major components. Houses action of the pistol through four major components. Controls functioning of the pistol.

e. **Magazine**: Holds cartridges in place for stripping and chambering.

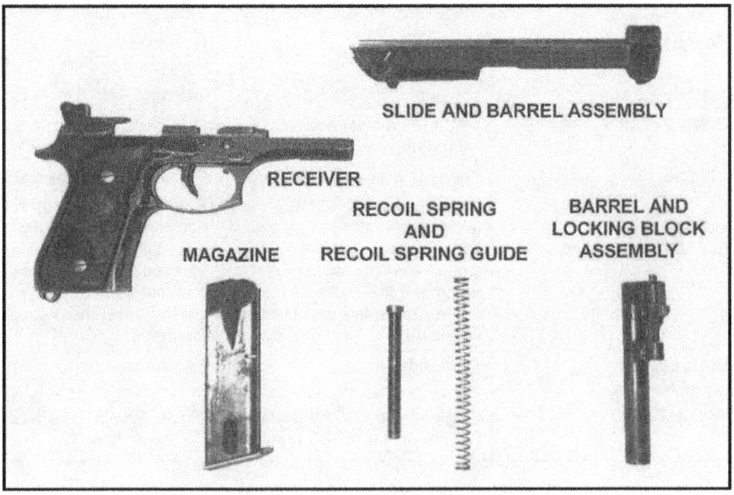

Figure 1-3. Major components, M9.

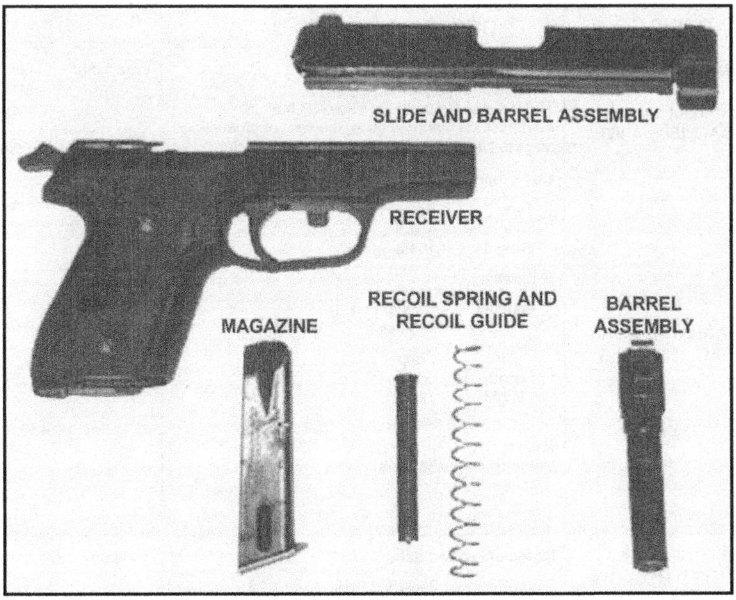

Figure 1-4. Major components, M11.

1-3. AMMUNITION

M9 and M11 pistols use several different types of 9-mm ammunition. Use only authorized ammunition manufactured to US and NATO specifications.

a. **Type and Characteristics**. The ammunition types (Table 1-2) and their characteristics follow:

(1) Cartridge, 9-mm ball, M882 with or without a cannelure.

(2) Cartridge, 9-mm dummy, M917.

> **WARNING**
>
> **Never fire heavily corroded or dented cartridges, cartridges with loose bullets, or any other rounds that you can see are defective. Failure to heed this warning could cause serious injury.**

CARTRIDGE	DESCRIPTION AND DATA	USE
9MM BALL, NATO M882 without cannelure DODAC 1305-A363 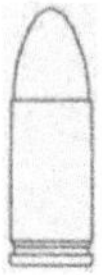 M882 with cannelure DODAC 1305-AA49 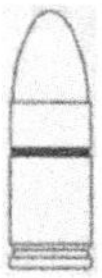	**Description:** Cartridge consists of a brass case and a copper projectile. The head of the cartridge case has the manufacturer's abbreviation, the year of production, and the NATO symbol (a cross inside a circle). **Complete round data:** Type Ball Length 1.165 in Weight 193 g **Projectile:** Material Copper jacket, Lead slug Color Copper Length 0.160 in Weight 124 g **Muzzle Velocity:** 385 m/s	**Weapons:** M9 and M11 Pistols Intended **Targets:** Personnel
9MM SUBSONIC JACKETED HOLLOW POINT MK 243 MOD 0 DODAC 1305-A260 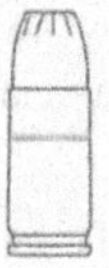	**Complete round data:** Type Jacketed Hollow Point Length 1.150 in Weight 210 g **Projectile:** Material Copper jacket, Lead slug Color Copper projectile Brass case Length Weight 147 g **Muzzle Velocity:** 975 ± 25 fps	Weapon: M11 only Users: US Army Criminal Investigation Command (USAICIDIC), which needs high lethality and limited overpenetration of the target to meet their mission profile.
M917 DUMMY ROUND 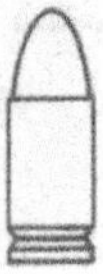	N/A	**Uses:** Dry fire and other training. **Misc:** Contains no propellant nor primer. Primer well remains open to prevent damage to the firing pin.

Table 1-2. Ammunition.

b. **Care, Handling, and Preservation.**

(1) Protect ammunition from mud, sand, and water. If the ammunition gets wet or dirty, wipe it off at once with a clean dry cloth. Wipe off light corrosion as soon as you find it. Turn in heavily corroded cartridges.

(2) Never expose ammunition to the direct rays of the sun. If the powder is hot, excessive pressure can develop when you fire the pistol.

(3) Never allow oil or grease to get on ammunition. Dust and other abrasives that collect on greasy ammunition can damage the operating parts of the pistol. Oiled cartridges produce excessive chamber pressure. Oil and lubricants can penetrate the cartridge primer and cause a misfire.

Section II. MAINTENANCE

Maintenance procedures include clearing, dissembling, inspecting, cleaning, lubricating, assembling, and checking the functioning of the M9 or M11 pistol.

1-4. CLEARING PROCEDURES, BOTH PISTOLS

The first step in maintenance is to clear the weapon. This applies in all situations, not just after firing. You must always assume that a weapon--any weapon, even yours--is loaded. To clear the pistol, perform the following procedures:

a. For the M9, place the decocking/safety lever in the SAFE (down) position. For the M11, thumb *down* the decocking lever.

b. Hold the pistol in the raised pistol position.

c. Depress the magazine release button and remove the magazine from the pistol.

d. Pull the slide to the rear and remove any chambered round.

e. Push the slide stop *up*, locking the slide to the rear.

f. Look into the chamber to ensure that it is empty.

1-5. GENERAL DISASSEMBLY OF M9 PISTOL

To disassemble the pistol, perform the following procedures:

a. Clear and unload the pistol.

b. Depress the slide stop and let the slide go forward.

c. Hold the pistol in your right hand with the muzzle slightly raised.

d. Press the disassembly lever button with your forefinger.

e. With your thumb, rotate the disassembly lever downward until it stops.

f. Pull the slide and barrel assembly forward and remove it from the receiver.

g. Carefully and lightly compress the recoil spring and spring guide. At the same time, lift up and remove them. Allow the recoil spring to stretch slowly.

h. Separate the recoil spring from the spring guide.

i. While pushing the barrel forward slightly, push in on the locking block plunger.

j. Lift and remove the locking block and barrel assembly from the slide.

1-6. GENERAL DISASSEMBLY OF M11 PISTOL

To disassemble the pistol--

a. Clear and unload the pistol.

b. Draw back the slide to the stop, and lock the slide in the open position by thumbing up the slide-catch lever.

c. Check visually and physically to ensure that the chamber is empty and that the magazine well is clear.

d. Thumb down the take-down lever.

e. Draw back the slide slightly to disengage the slide-catch lever.

f. Hold the slide firmly and allow it to glide forward gently.

g. Now slip the complete system (slide, barrel, recoil spring, and guide) forward and off the frame.

h. Remove the recoil spring and guide, taking care to control the release of spring tension, while at the same time directing the assembly away from your body.

i. Remove the barrel from the slide.

1-7. INSPECTION

Inspection begins with the pistol disassembled down to its major components. Shiny surfaces do not mean the parts are unserviceable. Inspect all surfaces for visible damage, cracks, burrs, and chips.

1-8. CLEANING, LUBRICATION, AND PREVENTIVE MAINTENANCE

Immediately after firing the M9 or M11 pistol, you should disassemble it into its major components and clean it. Clean all metal components and surfaces that have been exposed to powder fouling using CLP on a bore-cleaning patch. Use the same procedure to clean the receiver. After you have cleaned and wiped it dry, use a cloth and rub on a thin coat of CLP. This lubricates and preserves the exposed metal parts during all normal temperature ranges. When you are not using your pistol, inspect it weekly, and clean and lubricate it when necessary.

CAUTION

When using CLP, never use other cleaners. Never mix CLP with RBC or LSA.

a. Clear and disassemble the weapon.

b. Wipe or brush dirt, dust, and carbon buildup from the disassembled pistol.

c. Use CLP to help remove carbon buildup and stubborn dirt and grime.

d. Pay particular attention to the bolt face, guide rails on the receiver, grooves on the slide, and other hard-to-reach areas.

CAUTION

Never use mineral spirits, paint thinner, or dry cleaning solvent to clean the pistol. Use only issued lubricants and cleaners such as CLP or LSA.

e. Clean the bore and chamber using CLP, fresh swabs, and a bore brush.

f. Lubricate the pistol by covering all surfaces, including the bore and chamber, with a light coat of CLP. In extremely hot or cold weather, refer to the technical manual for lubricating procedures and materials.

1-9. M9 GENERAL ASSEMBLY

To assemble the M9 pistol, simply reverse the procedures used to disassemble the pistol:

a. Grasp the slide with the bottom facing up.

b. With the other hand, grasp the barrel assembly with the locking block facing up.

c. Insert the muzzle into the forward end of the slide and, at the same time, lower the rear of the barrel assembly by aligning the extractor cutout with the extractor.

NOTE: The locking block will fall into the locked position in the slide.

d. Insert the recoil spring onto the recoil spring guide.

CAUTION

Maintain spring tension until the spring guide seats fully in the cutaway on the locking block. Failure to maintain spring tension could cause injury to personnel, or the spring could become damaged or lost.

e. Insert the end of the recoil spring and the recoil spring guide into the recoil spring housing. At the same time, compress the recoil spring guide until it seats fully on the locking block cutaway.

WARNING

NEVER pull the trigger while placing the slide on the receiver. Doing this could damage the slide.

f. Ensure that the hammer is uncocked and that the firing pin block lever is in the *down* position. The decocking/safety lever should be in the SAFE position.

g. Grasp the slide and barrel assembly with the sights *up*, and align the slide on the receiver assembly guide rails.

h. Push until the rear of the slide moves a short distance beyond the rear of the receiver assembly, and then hold. At the same time, use the other hand to rotate the disassembly latch lever upward. A click indicates a positive lock.

1-10. M11 GENERAL ASSEMBLY

To assemble the M11 pistol, simply reverse the procedures used to disassemble the pistol:

a. Insert the barrel assembly into the slide assembly and fit the recoil spring and guide.

NOTE: If the recoil spring and guide were separated during disassembly, slip the narrow end of the spring over the guide first.

> **WARNING**
>
> **While compressing the spring, direct the assembly away from your body.**

b. While compressing the spring, direct the assembly away from your body.

c. Slip the assemble slide system into the frame until it contacts the end stop, and lock it with the slide-catch lever.

d. Thumb *up* the take-down lever.

e. Thumb *down* the slide-catch lever in order to close the slide.

f. Decock the hammer with the decocking lever.

1-11. M9 SAFETY AND FUNCTION CHECK

Always perform a function check after reassembling the pistol, to ensure it is working properly:

a. Clear the pistol in accordance with the unloading procedures.

b. Depress the slide stop, letting the slide go forward.

c. Insert an empty magazine into the pistol, and ensure that the magazine catch locks the magazine in place.

d. Retract the slide fully and release it. The slide should lock to the rear.

e. Depress the magazine release button and remove the magazine.

*f. Visually inspect to ensure the decocking/safety lever is in the SAFE (*down*) position.

*g. Depress the slide stop. When the slide goes forward, the hammer should fall to the full forward position. The decocking/safety lever must remain in the SAFE (*down*) position.

***NOTE:** If the safety lever moves to the fire position (*up*) during steps f. and g. above, then the weapon is non-mission capable. Evacuate the weapon to field-level maintenance for repair.

h. Squeeze and release the trigger. The firing pin block should move up and down and the hammer should not move.

i. Place the decocking/safety lever in the FIRE (*up*) position.

j. Squeeze the trigger to check double action. The hammer should cock and fall.

k. Squeeze the trigger again. Hold it to the rear. Manually retract and release the slide. Release the trigger. You should hear a click, and the hammer should not fall.

l. Squeeze the trigger to check single action. The hammer should fall.

1-12. M11 SAFETY AND FUNCTION CHECK

Always perform a function check after the pistol is reassembled to ensure it is working properly. To perform a function check:

a. Clear the pistol in accordance with the unloading procedures.

b. Depress the slide-catch lever.

c. Insert an empty magazine into the pistol, and ensure that the magazine catch locks the magazine in place.

d. Retract the slide and release it. The magazine follower should push up on the slide catch lever, locking the slide to the rear.

e. Depress the magazine catch button allowing the magazine to fall free.

f. Depress the slide-catch lever, allowing the slide to return fully forward.

g. Depress the decocking lever, allowing the hammer to fall.
h. Squeeze the trigger to check the double action. The hammer should cock and fall.
i. Squeeze the trigger again and hold it to the rear.
j. Manually retract and release the slide while holding the trigger to the rear.
k. Release the trigger; you should hear a click, but the hammer should not fall.
l. Squeeze the trigger to check the single action. The hammer should fall.

NOTE: If during the previous safety/function check the M11 performs as just described, then it is mission ready. If it fails to perform as described, sent it to unit maintenance or to the next authorized repair level.

Section III. OPERATION AND FUNCTION

This section provides detailed information on the operation and function of M9 and M11 pistols.

1-13. LOADING OF THE M9 PISTOL

Hold the pistol in the raised pistol position.

a. Insert the magazine into the magazine well of the pistol until you hear the magazine catch click. This ensures that the catch engaged properly.

b. With the pistol pointed in a safe direction, grasp the serrated portion of the slide, and retract the slide to the rear. Releasing the slide will strip a cartridge from the magazine and chamber it. The pistol is now loaded.

c. When ready to fire release the decocking/safety lever by rotating the decocking/safety lever to the fully upward position with the thumb.

WARNING

This pistol will fire in single or double action mode. If you pull the trigger with the decocking/safety lever in the fire (up) position and a round in the chamber, the pistol will fire from the hammer down, half or full-cocked position.

NOTE: For double-action fire, ensure the hammer is in the upward or half-cocked position. Squeezing the trigger cocks and releases the hammer.

d. After you fire the first round, the pistol will continue to fire in single-action mode. When the hammer is down, you can fire single action by manually cocking the hammer with your thumb. When you have fired the last round, the slide remains to the rear.

e. Always ensure that the muzzle is pointing in a safe direction, with your finger *off* the trigger.

f. Never try to load or unload any firearm inside a vehicle, building, or other confined space, except in a properly constructed shooting range or bullet trap. Enclosed areas offer few or no completely safe directions to point the firearm. For this reason, if the weapon discharges accidentally, the risk of injury or property damage is high.

g. Before loading, always clean excess grease and oil from the bore and chamber. Ensure that the barrel is clear of obstructions. The presence of any foreign matter in the barrel could bulge or burst the barrel, or cause other damage to the firearm or serious injury to the shooter or to others.

CHAPTER 2

PISTOL MARKSMANSHIP TRAINING

Marksmanship training is divided into two phases: preparatory marksmanship training and range firing. Each phase may be divided into separate instructional steps. All marksmanship training must be progressive. Combat marksmanship techniques should be practiced after the basics have been mastered.

Section I. BASIC MARKSMANSHIP

The main use of the pistol is to engage an enemy at close range with quick, accurate fire. Accurate shooting results from knowing and correctly applying the elements of marksmanship. The elements of combat pistol marksmanship are:

- Grip.
- Aiming.
- Breath control.
- Trigger squeeze.
- Target engagement.
- Positions.

2-1. GRIP

A proper grip is one of the most important fundamentals of quick fire. The weapon must become an extension of the hand and arm; it should replace the finger in pointing at an object. The firer must apply a firm, uniform grip to the weapon.

a. **One-Hand Grip**. Hold the weapon in the nonfiring hand; form a V with the thumb and forefinger of the strong hand (firing hand). Place the weapon in the V with the front and rear sights in line with the firing arm. Wrap the lower three fingers around the pistol grip, putting equal pressure with all three fingers to the rear. Allow the thumb of the firing hand to rest alongside the weapon without pressure (Figure 2-1). Grip the weapon tightly until the hand begins to tremble; relax until the trembling stops. At this point, the necessary pressure for a proper grip has been applied. Place the trigger finger on the trigger between the tip and second joint so that it can be squeezed to the rear. The trigger finger must work independently of the remaining fingers.

NOTE: If any of the three fingers on the grip are relaxed, the grip must be reapplied.

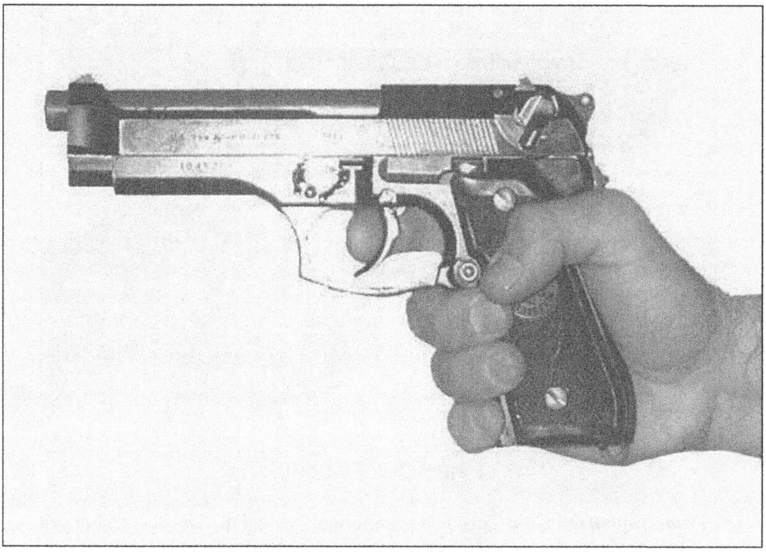

Figure 2-1. One-hand grip.

b. **Two-Hand Grip**. The two-hand grip allows the firer to steady the firing hand and provide maximum support during firing. The nonfiring hand becomes a support mechanism for the firing hand by wrapping the fingers of the nonfiring hand around the firing hand. Two-hand grips are recommended for all pistol firing.

WARNING

Do not place the nonfiring thumb in the rear of the weapon. The recoil upon firing could result in personal injury.

(1) ***Fist Grip***. Grip the weapon as with the one-hand grip. Firmly close the fingers of the nonfiring hand over the fingers of the firing hand, ensuring that the index finger from the nonfiring hand is between the middle finger of the firing hand and the trigger guard. Place the nonfiring thumb alongside the firing thumb (Figure 2-2).

NOTE: Depending upon the individual firer, he may chose to place the index finger of his nonfiring hand on the front of the trigger guard since M9 and M11 pistols have a recurved trigger guard designed for this purpose.

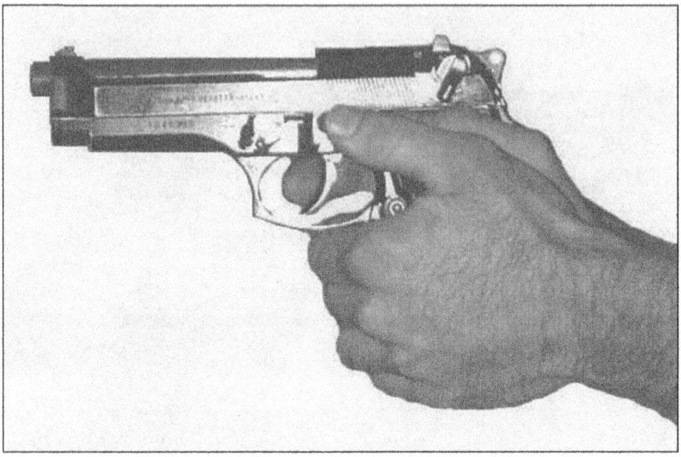

Figure 2-2. Fist grip.

(2) ***Palm-Supported Grip***. This grip is commonly called the cup and saucer grip. Grip the firing hand as with the one-hand grip. Place the nonfiring hand under the firing hand, wrapping the nonfiring fingers around the back of the firing hand. Place the nonfiring thumb over the middle finger of the firing hand (Figure 2-3).

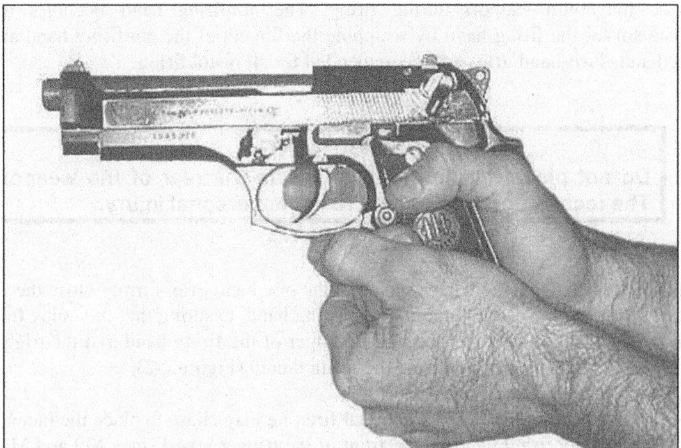

Figure 2-3. Palm-supported grip.

(3) ***Weaver grip***. Apply this grip the same as the fist grip. The only exception is that the nonfiring thumb is wrapped over the firing thumb (Figure 2-4).

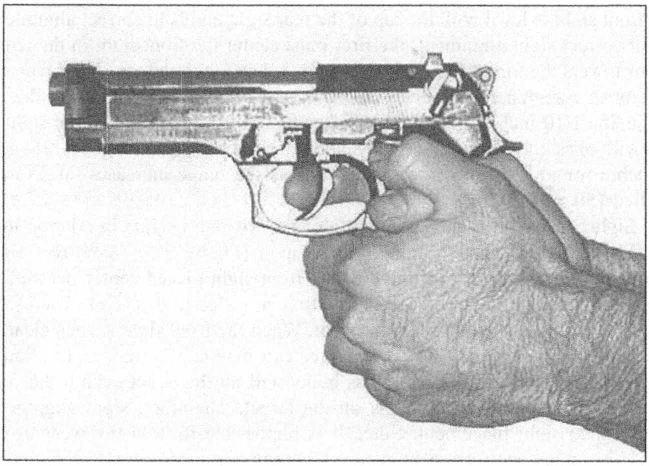

Figure 2-4. Weaver grip.

c. **Isometric Tension**. The firer raises his arms to a firing position and applies isometric tension. This is commonly known as the push-pull method for maintaining weapon stability. Isometric tension is when the firer applies forward pressure with the firing hand and pulls rearward with the nonfiring hand with equal pressure. This creates an isometric force but never so much to cause the firer to tremble. This steadies the weapon and reduces barrel rise from recoil. The supporting arm is bent with the elbow pulled downward. The firing arm is fully extended with the elbow and wrist locked. The firer must experiment to find the right amount of isometric tension to apply.

NOTE: The firing hand should exert the same pressure as the nonfiring hand. If it does not, a missed target could result.

d. **Natural Point of Aim**. The firer should check his grip for use of his natural point of aim. He grips the weapon and sights properly on a distant target. While maintaining his grip and stance, he closes his eyes for three to five seconds. He then opens his eyes and checks for proper sight picture. If the point of aim is disturbed, the firer adjusts his stance to compensate. If the sight alignment is disturbed, the firer adjusts his grip to compensate by removing the weapon from his hand and reapplying the grip. The firer repeats this process until the sight alignment and sight placement remain almost the same when he opens his eyes. With sufficient practice, this enables the firer to determine and use his natural point of aim, which is the most relaxed position for holding and firing the weapon.

2-2. AIMING

Aiming is sight alignment and sight placement (Figure 2-5).

a. Sight alignment is the centering of the front blade in the rear sight notch. The top of the front sight is level with the top of the rear sight and is in correct alignment with the eye. For correct sight alignment, the firer must center the front sight in the rear sight. He raises or lowers the top of the front sight so it is level with the top of the rear sight. Sight alignment is essential for accuracy because of the short sight radius of the pistol. For example, if a 1/10-inch error is made in aligning the front sight in the rear sight, the firer's bullet will miss the point of aim by about 15 inches at a range of 25 meters. The 1/10-inch error in sight alignment magnifies as the range increases--at 25 meters, it is magnified 150 times.

b. Sight placement is the positioning of the weapon's sights in relation to the target as seen by the firer when he aims the weapon (Figure 2-5). A correct sight picture consists of correct sight alignment with the front sight placed center mass of the target. The eye can focus on only one object at a time at different distances. Therefore, the last focus of the eye is always on the front sight. When the front sight is seen clearly, the rear sight and target will appear hazy. The firer can maintain correct sight alignment only through focusing on the front sight. His bullet will hit the target even if the sight picture is partly off center but still remains on the target. Therefore, sight alignment is more important than sight placement. Since it is impossible to hold the weapon completely still, the firer must apply trigger squeeze and maintain correct sight alignment while the weapon is moving in and around the center of the target. This natural movement of the weapon is referred to as wobble area. The firer must strive to control the limits of the wobble area through proper grip, breath control, trigger squeeze, and positioning.

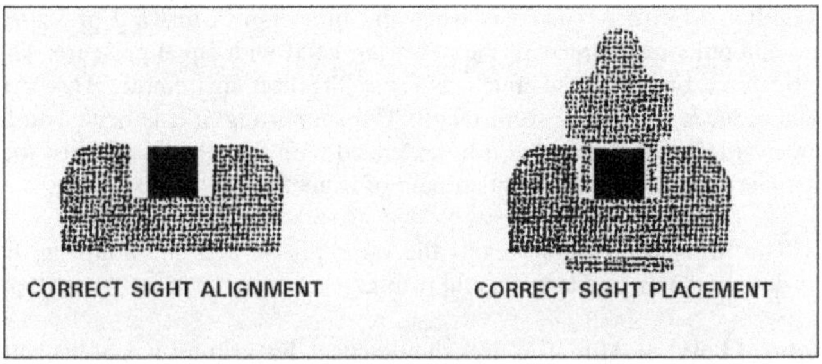

Figure 2-5. Correct sight alignment and sight picture.

c. Focusing on the front sight while applying proper trigger squeeze will help the firer resist the urge to jerk the trigger and anticipate the moment the weapon will fire. Mastery of trigger squeeze and sight alignment requires practice. Trainers should use concurrent training stations or have fire ranges to enhance proficiency of marksmanship skills.

2-3. BREATH CONTROL

To attain accuracy, the firer must learn to hold his breath properly at any time during the breathing cycle. This must be done while aiming and squeezing the trigger. While the procedure is simple, it requires explanation, demonstration, and supervised practice. To hold his breath properly, the firer takes a breath, lets it out, then inhales normally, lets a little out until comfortable, holds, and then fires. It is difficult to maintain a steady position keeping the front sight at a precise aiming point while breathing. Therefore, the firer should be taught to inhale, then exhale normally, and hold his breath at the moment of the natural respiratory pause (Figure 2-6). Breath control, firing at a single target.) The shot must then be fired before he feels any discomfort from not breathing. When multiple targets are presented, the firer must learn to hold his breath at any part of the breathing cycle (Figure 2-7). Breath control must be practiced during dry-fire exercises until it becomes a natural part of the firing process.

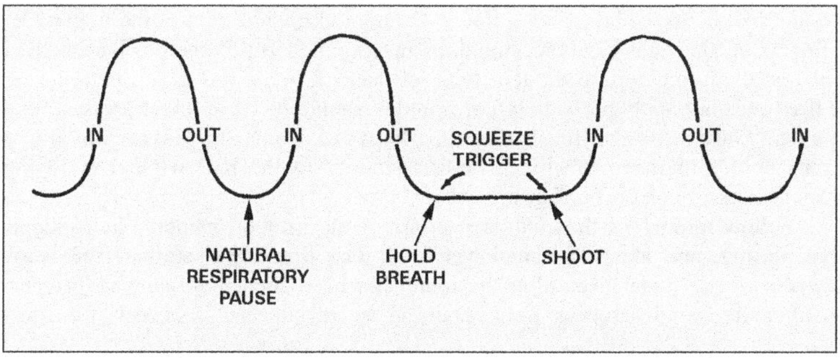

Figure 2-6. Breath control, firing at a single target.

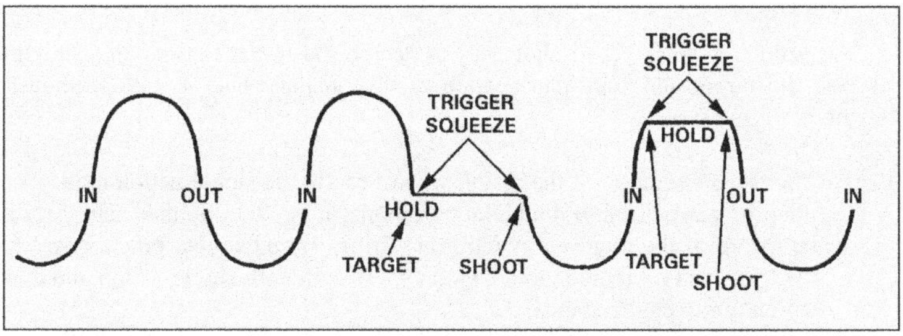

Figure 2-7. Breath control, firing at timed or multiple targets.

2-4. TRIGGER SQUEEZE

Improper trigger squeeze causes more misses than any other step of preparatory marksmanship. Poor shooting is caused by the aim being disturbed before the bullet leaves the barrel of the weapon. This is usually the result of the firer jerking the trigger or flinching. A slight off-center pressure of the trigger finger on the trigger can cause the weapon to move and disturb the firer's sight alignment. Flinching is an automatic human reflex caused by anticipating the recoil of the weapon. Jerking is an effort to fire the weapon at the precise time the sights align with the target. For more on problems in target engagement, see paragraph 2-5.

a. Trigger squeeze is the independent movement of the trigger finger in applying increasing pressure on the trigger straight to the rear, without disturbing the sight alignment until the weapon fires. The trigger slack, or free play, is taken up first, and the squeeze is continued steadily until the hammer falls. If the trigger is squeezed properly, the firer will not know exactly when the hammer will fall; thus, he will not tend to flinch or heel, resulting in a bad shot. Novice firers must be trained to overcome the urge to anticipate recoil. Proper application of the fundamentals will lower this tendency.

b. To apply correct trigger squeeze, the trigger finger should contact the trigger between the tip of the finger and the second joint (without touching the weapon anywhere else). Where contact is made depends on the length of the firer's trigger finger. If pressure from the trigger finger is applied to the right side of the trigger or weapon, the strike of the bullet will be to the left. This is due to the normal hinge action of the fingers. When the fingers on the right hand are closed, as in gripping, they hinge or pivot to the left, thereby applying pressure to the left (with left-handed firers, this action is to the right). The firer must not apply pressure left or right but should increase finger pressure straight to the rear. Only the trigger finger should perform this action. Dry-fire training improves a firer's ability to move the trigger finger straight to the rear without cramping or increasing pressure on the hand grip.

c. Follow-through is the continued effort of the firer to maintain sight alignment before, during, and after the round has fired. The firer must continue the rearward movement of the finger even after the round has been fired. Releasing the trigger too soon after the round has been fired results in an uncontrolled shot, causing a missed target.

(1) The firer who is a good shot holds the sights of the weapon as nearly on the target center as possible and continues to squeeze the trigger with increasing pressure until the weapon fires.

(2) The soldier who is a bad shot tries to "catch his target" as his sight alignment moves past the target and fires the weapon at that instant. This is called ambushing, which causes trigger jerk.

NOTE: The trigger squeeze of the pistol, when fired in the single-action mode, is 5.50 pounds; when fired in double-action mode, it is 12.33 pounds. The firer must be aware of the mode in which he is firing. He must also practice squeezing the trigger in each mode to develop expertise in both single-action and double-action target engagements.

2-5. TARGET ENGAGEMENT

To engage a single target, the firer applies the method discussed in paragraph 2-4. When engaging multiple targets in combat, he engages the closest and most dangerous multiple target first and fires at it with two rounds. This is called controlled pairs. The firer then traverses and acquires the next target, aligns the sights in the center of mass, focuses on the front sight, applies trigger squeeze, and fires. He ensures his firing arm elbow and wrist are locked during all engagements. If he has missed the first target and has fired upon the second target, he shifts back to the first and engages it. Some problems in target engagement are as follows:

a. **Recoil Anticipation**. When a soldier first learns to shoot, he may begin to anticipate recoil. This reaction may cause him to tighten his muscles during or just before the hammer falls. He may fight the recoil by pushing the weapon downward in anticipating or reacting to its firing. In either case, the rounds will not hit the point of aim. A good method to show the firer that he is anticipating the recoil is the ball-and-dummy method (see paragraph 2-14).

b. **Trigger Jerk**. Trigger jerk occurs when the soldier sees that he has acquired a good sight picture at center mass and "snaps" off a round before the good sight picture is lost. This may become a problem, especially when the soldier is learning to use a flash sight picture (see paragraph 2-7b).

c. **Heeling**. Heeling is caused by a firer tightening the large muscle in the heel of the hand to keep from jerking the trigger. A firer who has had problems with jerking the trigger tries to correct the fault by tightening the bottom of the hand, which results in a heeled shot. Heeling causes the strike of the bullet to hit high on the firing hand side of the target. The firer can correct shooting errors by knowing and applying correct trigger squeeze.

2-6. POSITIONS

The qualification course is fired from a standing, kneeling, or crouch position. During qualification and combat firing, soldiers must practice all of the firing positions described below so they become natural movements. Though these positions seem natural, practice sessions must be conducted to ensure the habitual attainment of correct firing positions. Practice in assuming correct firing positions ensures that soldiers can quickly assume these positions without a conscious effort. Pistol marksmanship requires a soldier to rapidly apply all the fundamentals at dangerously close targets while under stress. Assuming a proper position to allow for a steady aim is critical to survival.

NOTE: During combat, there may not be time for a soldier to assume a position that will allow him to establish his natural point of aim. Firing from a covered position may require the soldier to adapt his shooting stance to available cover.

a. **Pistol-Ready Position**. In the pistol-ready position, hold the weapon in the one-hand grip. Hold the upper arm close to the body and the forearm at about a 45-degree angle. Point the weapon toward target center as you move forward (Figure 2-8).

Figure 2-8. Pistol-ready position.

b. **Standing Position without Support**. Face the target (Figure 2-9). Place feet a comfortable distance apart, about shoulder width. Extend the firing arm and attain a two-hand grip. The wrist and elbow of the firing arm are locked and pointed toward target center. Keep the body straight with the shoulders slightly forward of the buttocks.

Figure 2-9. Standing position without support.

c. **Kneeling Position**. In the kneeling position, ground only your firing-side knee as the main support (Figure 2-10). Vertically place your firing-side foot, used as the main support, under your buttocks. Rest your body weight on the heel and toes. Rest your nonfiring arm just above the elbow on the knee not used as the main body support. Use the two-handed grip for firing. Extend the firing arm, and lock the firing-arm elbow and wrist to ensure solid arm control.

Figure 2-10. Kneeling position.

d. **Crouch Position**. Use the crouch position when surprise targets are engaged at close range (Figure 2-11). Place the body in a forward crouch (boxer's stance) with the knees bent slightly and trunk bent forward from the hips to give faster recovery from recoil. Place the feet naturally in a position that allows another step toward the target. Extend the weapon straight toward the target, and lock the wrist and elbow of the firing arm. It is important to consistently train with this position, since the body will automatically crouch under conditions of stress such as combat. It is also a faster position from which to change direction of fire.

Figure 2-11. Crouch position.

e. **Prone Position**. Lie flat on the ground, facing the target (Figure 2-12). Extend your arms in front with the firing arm locked. (Your arms may have to be slightly unlocked for firing at high targets.) Rest the butt of the weapon on the ground for single, well-aimed shots. Wrap the fingers of the nonfiring hand around the fingers of the firing hand. Face forward. Keep your head down between your arms and behind the weapon as much as possible.

Figure 2-12. Prone position.

f. **Standing Position with Support**. Use available cover for support--for example, a tree or wall to stand behind (Figure 2-13). Stand behind a barricade with the firing side on line with the edge of the barricade. Place the knuckles of the nonfiring fist at eye level against the edge of the barricade. Lock the elbow and wrist of the firing arm. Move the foot on the nonfiring side forward until the toe of the boot touches the bottom of the barricade.

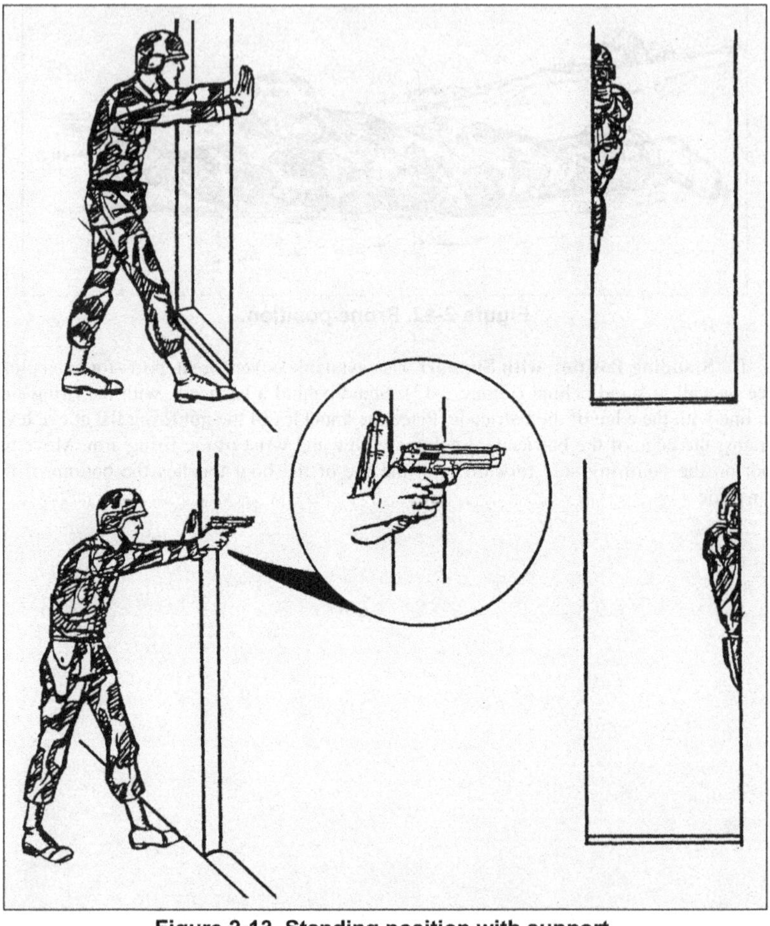

Figure 2-13. Standing position with support.

g. **Kneeling Supported Position**. Use available cover for support--for example, use a low wall, rocks, or vehicle (Figure 2-14). Place your firing-side knee on the ground. Bend the other knee and place the foot (nonfiring side) flat on the ground, pointing toward the target. Extend arms alongside and brace them against available cover. Lock the wrist and elbow of your firing arm. Place the nonfiring hand around the fist to support the firing arm. Rest the nonfiring arm just above the elbow on the nonfiring-side knee.

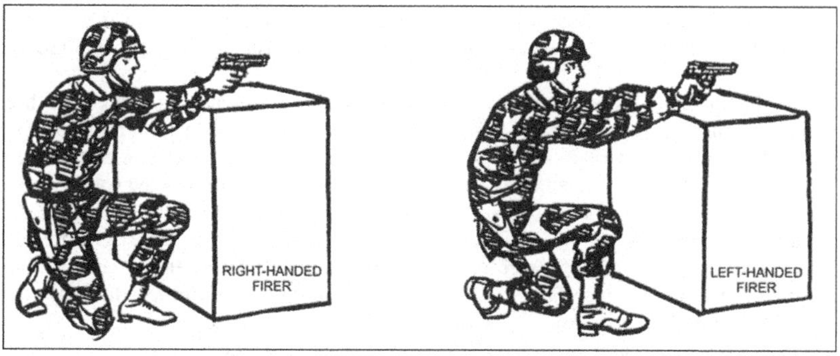

Figure 2-14. Kneeling supported.

Section II. COMBAT MARKSMANSHIP

After a soldier becomes proficient in the fundamentals of marksmanship, he progresses to advanced techniques of combat marksmanship. The main use of the pistol is to engage the enemy at close range with quick, accurate fire. In shooting encounters, it is not the first round fired that wins the engagement, but the first accurately fired round. The soldier should use his sights when engaging the enemy unless this would place the weapon within arm's reach of the enemy.

2-7. TECHNIQUES OF FIRING

Firing techniques include the use of hand-and-eye coordination, flash sight picture, quick-fire point shooting, and quick-fire sighting.

a. **Hand-and-Eye Coordination**. Hand-and-eye coordination is not a natural, instinctive ability for all soldiers. It is usually a learned skill obtained by practicing the use of a flash sight picture (see paragraph b below). The more a soldier practices raising the weapon to eye level and obtaining a flash sight picture, the more natural the relationship between soldier, sights, and target becomes. Eventually, proficiency elevates to a point so that the soldier can accurately engage targets in the dark. Each soldier must be aware of this trait and learn how to use it best. Poorly coordinated soldiers can achieve proficiency through close supervision from their trainers. Everyone has the ability to point at an object. Since pointing the forefinger at an object and extending the weapon toward a target are much the same, the combination of the two are natural. Making the soldier aware of this ability and teaching him how to apply it results in success when engaging enemy targets in combat.

(1) The eyes focus instinctively on the center of any object observed. After the object is sighted, the firer aligns his sights on the center of mass, focuses on the front sight, and applies proper trigger squeeze. Most crippling or killing hits result from maintaining the focus on the center of mass. The eyes must remain fixed on some part of the target throughout firing.

(2) When a soldier points, he instinctively points at the feature on the object on which his eyes are focused. An impulse from the brain causes the arm and hand to stop when the finger reaches the proper position. When the eyes are shifted to a new object or feature,

the finger, hand, and arm also shift to this point. It is this inherent trait that can be used by the soldier to engage targets rapidly and accurately. This instinct is called hand-and-eye coordination.

b. **Flash Sight Picture**. Usually, when engaging an enemy at pistol range, the firer has little time to ensure a correct sight picture. The quick-kill (or natural point of aim) method does not always ensure a first-round hit. A compromise between a correct sight picture and the quick-kill method is known as a flash sight picture. As the soldier raises the weapon to eye level, his point of focus switches from the enemy to the front sight, ensuring that the front and rear sights are in proper alignment left and right, but not necessarily up and down. Pressure is applied to the trigger as the front sight is being acquired, and the hammer falls as the flash sight picture is confirmed. Initially, this method should be practiced slowly, with speed gained as proficiency increases.

c. **Quick-Fire Point Shooting**. This is for engaging an enemy at less than 5 yards and is also useful for night firing. Using a two-hand grip, the firer brings the weapon up close to the body until it reaches chin level. He then thrusts it forward until both arms are straight. The arms and body form a triangle, which can be aimed as a unit. In thrusting the weapon forward, the firer can imagine that there is a box between him and the enemy, and he is thrusting the weapon into the box. The trigger is smoothly squeezed to the rear as the elbows straighten.

d. **Quick-Fire Sighting**. This technique is for engaging an enemy at 5 to 10 yards away and only when there is no time available to get a full picture. The firing position is the same as for quick-fire point shooting. The sights are aligned left and right to save time, but not up and down. The firer must determine in practice what the sight picture will look like and where the front sight must be aimed to hit the enemy in the chest.

2-8. TARGET ENGAGEMENT

In close combat, there is seldom time to precisely apply all of the fundamentals of marksmanship. When a soldier fires a round at the enemy, he often does not know if he hits his target. Therefore, two rounds should be fired at the target. This is called controlled pairs. If the enemy continues to attack, two more shots should be placed in the pelvic area to break the body's support structure, causing the enemy to fall.

2-9. TRAVERSING

In close combat, the enemy may be attacking from all sides. The soldier may not have time to constantly change his position to adapt to new situations. The purpose of the crouching or kneeling 360-degree traverse is to fire in any direction without moving the feet.

a. **Crouching 360-Degree Traverse**. The following instructions are for a right-handed firer. The two-hand grip is used at all times except for over the right shoulder. The firer remains in the crouch position with feet almost parallel to each other. Turning will be natural on the balls of the feet.

(1) ***Over the Left Shoulder*** (Figure 2-15): The upper body is turned to the left, the weapon points to the left rear with the elbows of both arms bent. The left elbow is naturally bent more than the right elbow.

(2) ***Traversing to the Left*** (Figure 2-16): The upper body turns to the right, and the right firing arm straightens out. The left arm is slightly bent.

(3) ***Traversing to the Front*** (Figure 2-17): The upper body turns to the front as the left arm straightens out. Both arms are straight forward.

(4) ***Traversing to the Right*** (Figure 2-18): The upper body turns to the right as both elbows bend. The right elbow is naturally bent more than the left.

Figure 2-15. Traversing over the left shoulder.

Figure 2-16. Traversing to the left.

Figure 2-17. Traversing to the front.

Figure 2-18. Traversing to the right.

(5) ***Traversing to the Right Rear*** (Figure 2-19): The upper body continues to turn to the right until it reaches a point where it cannot go further comfortably. Eventually the left hand must be released from the fist grip, and the firer will be shooting to the right rear with the right hand.

Figure 2-19. Traversing to the right rear.

b. **Kneeling 360-Degree Traverse**. The following instructions are for right-handed firers. The hands are in a two-hand grip at all times. The unsupported kneeling position is used. The rear foot must be positioned to the left of the front foot.

(1) ***Traversing to the Left Side*** (Figure 2-20): The upper body turns to a comfortable position toward the left. The weapon is aimed to the left. Both elbows are bent with the left elbow naturally bent more than the right elbow.

(2) ***Traversing to the Front*** (Figure 2-21): The upper body turns to the front, and a standard unsupported kneeling position is assumed. The right firing arm is straight, and the left elbow is slightly bent.

(3) ***Traversing to the Right Side*** (Figure 2-22): The upper body turns to the right as both arms straighten out.

(4) ***Traversing to the Rear*** (Figure 2-23): The upper body continues to turn to the right as the left knee is turned to the right and placed on the ground. The right knee is lifted off the ground and becomes the forward knee. The right arm is straight, while the left arm is bent. The direction of the kneeling position has been reversed.

Figure 2-20. Traversing to the left, kneeling.

Figure 2-21. Traversing to the front, kneeling.

Figure 2-22. Traversing to the right, kneeling.

Figure 2-23. Traversing to the rear, kneeling.

(5) ***Traversing to the New Right Side*** (Figure 2-24): The upper body continues to the right. Both elbows are straight until the body reaches a point where it cannot go further comfortably. Eventually, the left hand must be released from the fist grip, and the firer is shooting to the right with the one-hand grip.

Figure 2-24. Traversing to the new right side, kneeling.

c. **Training Method**. This method can be trained and practiced anywhere and, with the firer simulating a two-hand grip, without a weapon. The firer should be familiar with firing in all five directions.

2-10. COMBAT RELOADING TECHNIQUES

Overlooked as a problem for many years, reloading has resulted in many casualties due to soldiers' hands shaking or errors such as dropped magazines, magazines placed in the pistol backwards, or empty magazines placed back into the weapon. The stress state induced by a life-threatening situation causes soldiers to do things they would not otherwise do. Consistent, repeated training is needed to avoid such mistakes.

NOTE: These procedures should be used only in combat, not on firing ranges.

a. Develop a consistent method for carrying magazines in the ammunition pouches. All magazines should face down with the bullets facing forward and to the center of the body.

b. Know when to reload. When possible, count the number of rounds fired. However, it is possible to lose count in close combat. If this happens, there is a distinct difference in recoil of the pistol when the last round has been fired. Change magazines when two rounds may be left--one in the magazine and one in the chamber. This prevents being caught with an empty weapon at a crucial time. Reloading is faster with a round in the chamber since time is not needed to release the slide.

c. Obtain a firm grip on the magazine. This precludes the magazine being dropped or difficulty in getting the magazine into the weapon. Ensure the knuckles of the hand are toward the body while gripping as much of the magazine as possible. Place the index finger high on the front of the magazine when withdrawing from the pouch. Use the index finger to guide the magazine into the magazine well.

d. Know which reloading procedure to use for the tactical situation. There are three systems of reloading: rapid, tactical, and one-handed. Rapid reloading is used when the soldier's life is in immediate danger and the reload must be accomplished quickly. Tactical reloading is used when there is more time and it is desirable to keep the replaced magazine because there are rounds still in it or it will be needed again. One-handed reloading is used when there is an arm injury.

(1) ***Rapid Reloading***.

(a) Place your hand on the next magazine in the ammunition pouch to ensure there is another magazine.

(b) Withdraw the magazine from the pouch while releasing the other magazine from the weapon. Let the replaced magazine drop to the ground.

(c) Insert the replacement magazine, guiding it into the magazine well with the index finger.

(d) Release the slide, if necessary.

(e) Pick up the dropped magazine if time allows. Place it in your pocket, not back into the ammunition pouch where it may become mixed with full magazines.

(2) ***Tactical Reloading***.

(a) Place your hand on the next magazine in the ammunition pouch to ensure there is a remaining magazine.

(b) Withdraw the magazine from the pouch.

(c) Drop the used magazine into the palm of the nonfiring hand, which is the same hand holding the replacement magazine.

(d) Insert the replacement magazine, guiding it into the magazine well with the index finger.

(e) Release the slide, if necessary.

(f) Place the used magazine into a pocket. Do not mix it with full magazines.

(3) ***One-Hand Reloading, Right Hand***.

(a) Push the magazine release button with the thumb.

(b) Place the safety ON with the thumb if the slide is forward.

(c) Place the weapon backwards into the holster.

NOTE: If placing the weapon in the holster backwards is a problem, place the weapon between the calf and thigh to hold the weapon.

(d) Insert the replacement magazine.

(e) Withdraw the weapon from the holster.

(f) Remove the safety with the thumb if the slide is forward, or push the slide release if the slide is back.

(4) ***One-Hand Reloading, Left Hand***.

(a) Push the magazine release button with the middle finger.

(b) Place the weapon backwards into the holster.

NOTE: If placing the weapon in the holster backwards is a problem, place the weapon between the calf and thigh to hold the weapon.

(c) Insert the replacement magazine.

(d) Remove the weapon from the holster.

(e) Remove the safety with the thumb if the slide is forward, or push the slide release lever with the middle finger if the slide is back.

2-11. POOR VISIBILITY FIRING

Poor visibility firing with any weapon is difficult since shadows can be misleading to the firer. This is mainly true during EENT and EMNT (a half hour before dark and a half hour before dawn). Even though the pistol is a short-range weapon, the hours of darkness and poor visibility further decrease its effect. To compensate, the firer must use the three principles of night vision:

a. **Dark Adaptation**. This process conditions the eyes to see during poor visibility conditions. The eyes usually need about 30 minutes to become 90 percent adapted in a totally darkened area.

b. **Off-Center Vision**. When looking at an object in daylight, a person looks directly at it. However, at night he would see the object only for a few seconds. To see an object in darkness, he must concentrate on it while looking 6 to 10 degrees away from it.

c. **Scanning**. This is the short, abrupt, irregular movement of the firer's eyes around an object or area every 4 to 10 seconds. With artificial illumination, the firer uses night-fire techniques to engage targets, since targets seem to shift without moving.

NOTE: For more detailed information on the three principles of night vision, see FM 21-75.

2-12. CHEMICAL, BIOLOGICAL, RADIOLOGICAL, OR NUCLEAR

When firing a pistol under CBRN conditions, the firer should use optical inserts, if applicable. Firing in MOPP levels 1 through 3 should not be a problem for the firer. Unlike with a rifle, the firer acquires a sight picture with a pistol the same with or without a protective mask. MOPP4 is the only level that might present a problem for a firer, because that level requires him to wear gloves. Gloves could force him to adjust for proper grip and trigger squeeze. Firers should practice firing in MOPP4 to become proficient in CBRN firing.

Section III. COACHING AND TRAINING AIDS

Throughout preparatory marksmanship training, the coach-and-pupil method of training should be used. The proficiency of a pupil depends on how well the coach performs his duties. This section provides detailed information on coaching techniques and training aids for pistol marksmanship.

2-13. COACHING

The coach assists the firer by correcting errors, ensuring he takes proper firing positions, and ensuring he observes all safety precautions. The criteria for selecting coaches are a command responsibility. Coaches must have more experience in pistol marksmanship than the student firer. Duties of the coach during instructional practice and record fire include the following:

a. Checking that the--

- Weapon is clear.
- Ammunition is clean.
- Magazines are clean and operational.

b. Observing the firer to see that he--
- Takes the correct firing position.
- Loads the weapon properly and only on command.
- Takes up the trigger slack correctly.
- Squeezes the trigger correctly.
- Calls the shot each time he fires, except during quick fire and rapid fire.
- Holds his breath correctly.
- When he does not fire for 5 or 6 seconds, lowers the weapon and rests his arm.

c. Having the firer breathe deeply several times to relax if he is tense.

2-14. BALL-AND-DUMMY METHOD

In this method, the coach loads the weapon for the firer. He may hand the firer a loaded weapon or an empty one. When firing the empty weapon, the firer observes that in anticipating recoil he is forcing the weapon downward as the hammer falls. Repetition of the ball-and-dummy method helps reduce recoil anticipation.

2-15. CALLING OF THE SHOT

To call the shot is to state where the bullet should strike the target according to the sight picture at the instant the weapon fires, for example, "High," "a little low," "to the left," or "bull's eye." Another method of calling the shot is the clock system, for example, "three-ring hit at 8 o'clock" or "four-ring hit at 5 o'clock." Another method is to place a firing center beside the firer on the firing line. As soon as the shot is fired, the firer must place a finger on the target face or center where he expects the round to hit on the target. This method avoids guessing and computing for the firer. The immediate placing of the finger on the target face gives an accurate call. If the firer calls his shot incorrectly in range fire, he is failing to concentrate on sight alignment and trigger squeeze. Thus, as the weapon fires, he does not know what his sight picture is.

2-16. SLOW-FIRE EXERCISE

The slow-fire exercise is one of the most important exercises for both amateur and competitive marksmen. Coaches should ensure firers practice this exercise as much as possible. This is a dry-fire exercise.

a. To perform the slow-fire exercise, the firer assumes the standing position with the weapon pointed at the target. The firer should begin by using the two-hand grip, progressing to the one-hand grip as his skill increases. He takes in a normal breath and lets out part of it, locking the remainder in his lungs by closing his throat. He then relaxes, aims at the target, and takes the correct sight alignment and sight picture. He takes up the trigger slack and squeezes the trigger straight to the rear with steady, increasing pressure until the hammer falls, simulating firing.

b. If the firer does not cause the hammer to fall in 5 or 6 seconds, he should return to the pistol-ready position and rest his arm and hand. He then starts the procedure again. The action sequence that makes up this process can be summed up by the key word BRASS. It is a word the firer should think of each time he fires his weapon.

Breathe	Take a normal breath, let part of it out, and lock the remainder in the lungs by closing the throat.
Relax	Relax the body muscles.
Aim	Take correct sight alignment and sight picture, and focus the eye at the top of the front sight.
Slack	Take up the trigger slack.
Squeeze	Squeeze the trigger straight to the rear with steadily increasing pressure without disturbing sight alignment until the hammer falls.

c. Coaches should observe the front sight for erratic movements during the application of trigger squeeze. Proper application of trigger squeeze allows the hammer to fall without the front sight moving. A small bouncing movement of the front sight is acceptable. Firers should call the shot by the direction of movement of the front sight (high, low, left, or right).

2-17. AIR-OPERATED PISTOL, .177 MM

The air-operated pistol is used as a training device to teach the soldier the method of quick fire, to increase confidence in his ability, and to afford him more practice firing. A range can be set up almost anywhere with a minimum of effort and coordination, which is ideal for USAR and NG. If conducted on a standard range, live firing of pistols can be conducted along with the firing of the .177-mm air-operated pistol. Due to light recoil and little noise of the pistol, the soldier can concentrate on fundamentals. This helps build confidence because the soldier can hit a target faster and more accurately. The air-operated pistol should receive the same respect as any firearm. A thorough explanation of the weapon and a safety briefing are given to each soldier.

2-18. QUICK-FIRE TARGET TRAINING DEVICE

The QTTD (Figures 2-25 and 2-26) is used with the .177-mm air-operated pistol.

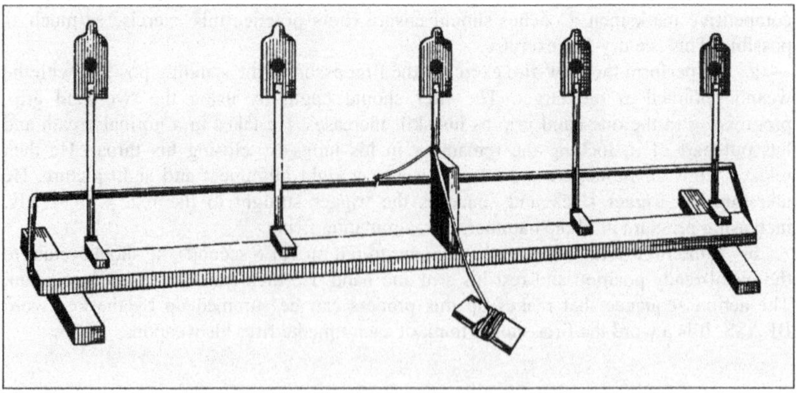

Figure 2-25. The quick-fire target training device.

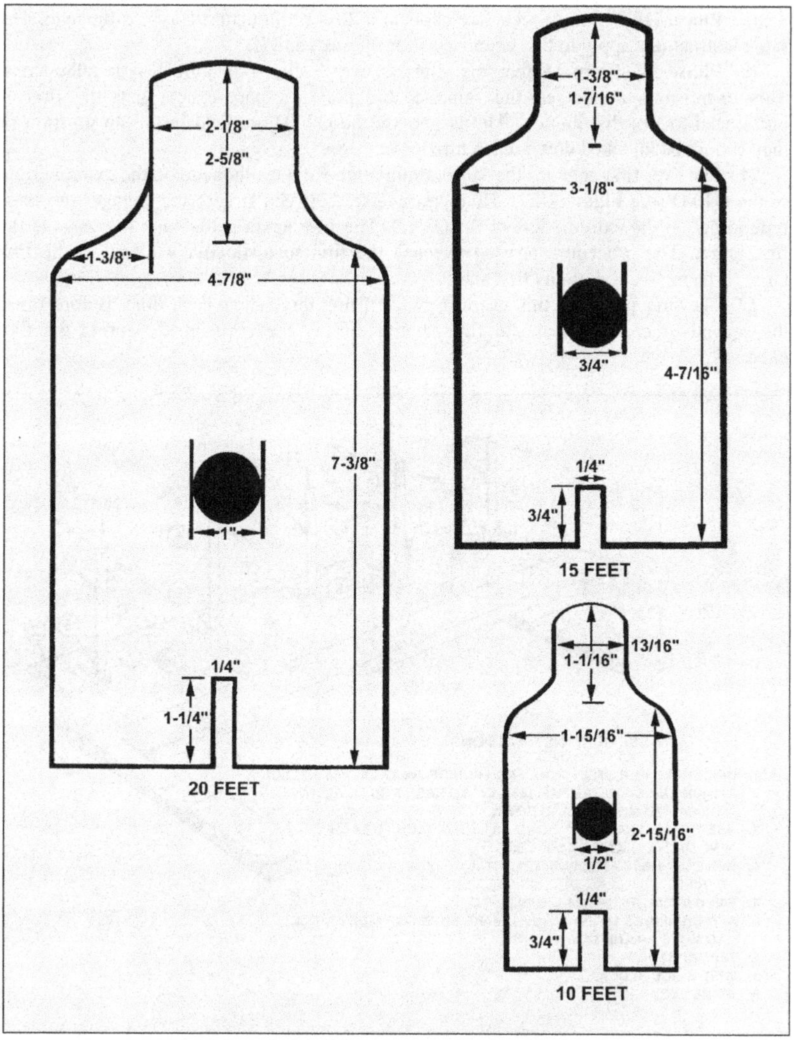

Figure 2-26. Dimensions for the QTTD.

a. **Phase I**. From 10 feet, five shots at a 20-foot miniature E-type silhouette. After firing each shot, the firer and coach discuss the results and make corrections.

b. **Phase II**. From 15 feet, five shots at a 20-foot miniature E-type silhouette. The same instructions apply to this exercise as for Phase I.

c. **Phase III**. From 20 feet, five shots at a 20-foot miniature E-type silhouette. The same instructions apply to this exercise as for Phases I and II.

d. **Phase IV**. From 15 feet, six shots at two 20-foot miniature E-type silhouettes. This exercise is conducted the same as the previous one, except that the firer is introduced to fire distribution. The targets on the QTTD are held in the up position so they cannot be knocked down when hit.

(1) The firer first engages the 20-foot miniature E-type silhouette on the extreme right of the QTTD (see Figure 2-27). He then traverses between targets and engages the same type target on the extreme left of the QTTD. The firer again shifts back to reengage the first target. The procedure is used to teach the firer to instinctively return to the first target if he misses it with his first shot.

(2) The firer performs this exercise twice, firing three shots each time. Before firing the second time, the coach and firer should discuss the errors made during the first exercise.

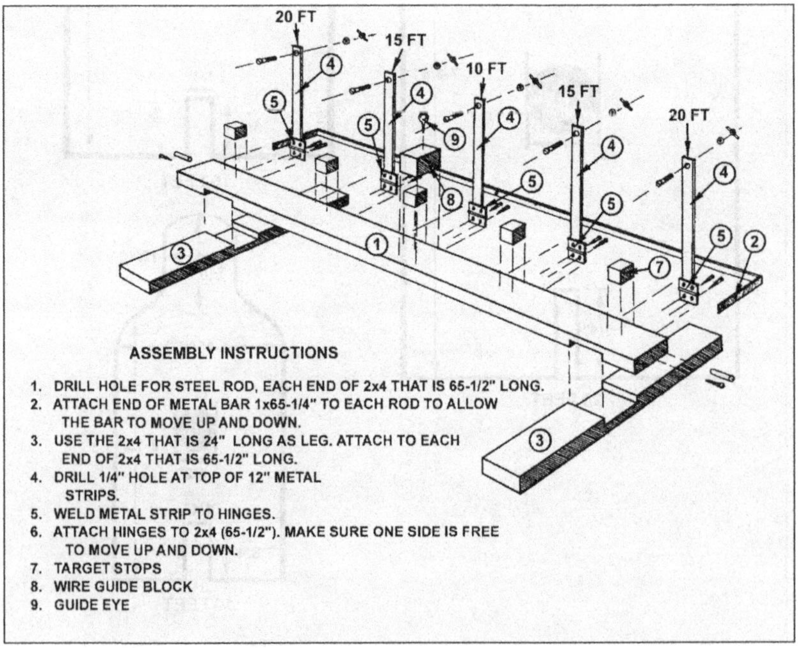

Figure 2-27. Miniature E-type silhouette for use with QTTD.

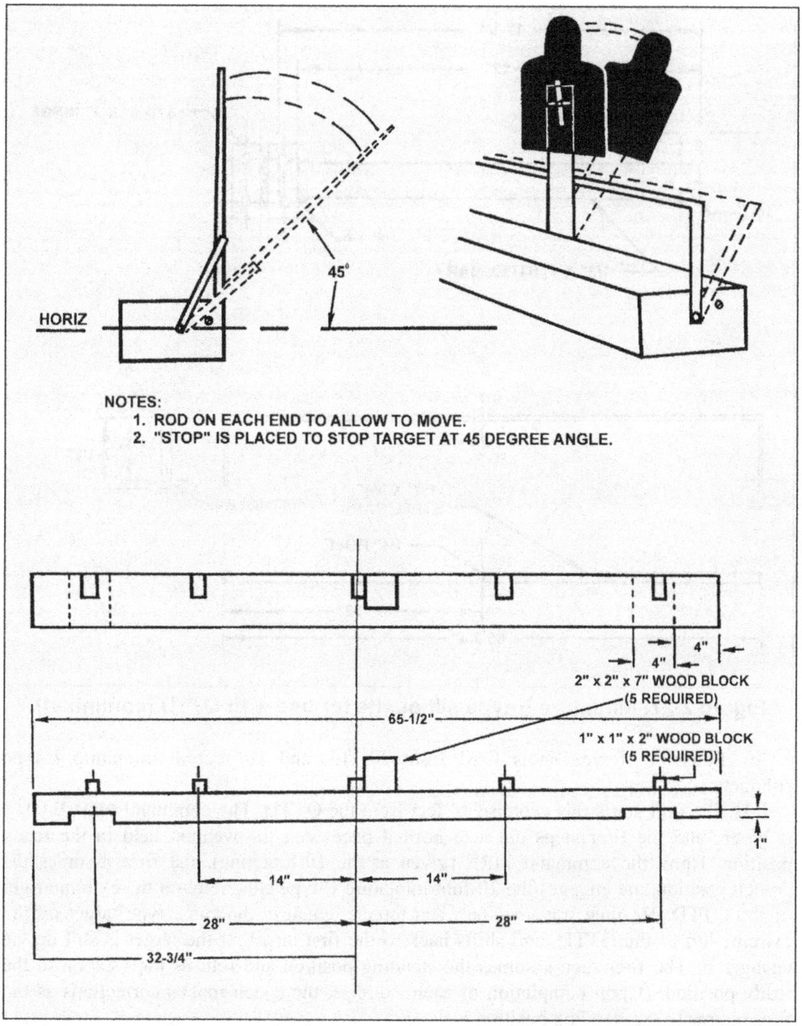

Figure 2-27. Miniature E-type silhouette for use with QTTD (continued).

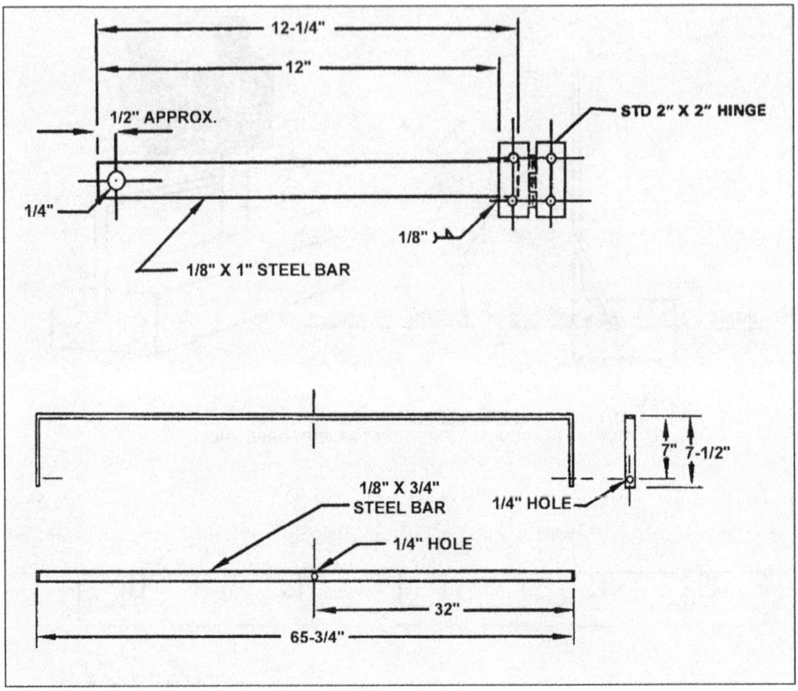

Figure 2-27. Miniature E-type silhouette for use with QTTD (continued).

e. **Phase V**. Seven shots fired from 20, 15, and 10 feet at miniature E-type silhouettes.

(1) The firer starts this exercise 30 feet from the QTTD. The command MOVE OUT is given, and the firer steps out at a normal pace with the weapon held in the ready position. Upon the command FIRE (given at the 20-foot line), the firer assumes the crouch position and engages the 20-foot miniature E-type silhouette on the extreme right of the QTTD. He then traverses between targets, engages the same type target on the extreme left of the QTTD, and shifts back to the first target. If the target is still up, he engages it. The firer then assumes the standing position and returns the weapon to the ready position. (Upon completion of each exercise, the coach makes corrections as the firer returns to the standing position.)

(2) On the command MOVE OUT, the firer again steps off at a normal pace. Upon the command FIRE (given at the 15-foot line), he engages the 15-foot targets on the QTTD. The same sequence of fire distribution is followed as with the previous exercise.

(3) During this exercise, the firer moves forward on command until he reaches the 10-foot line. At the command FIRE, the firer engages the 10-foot miniature E-type silhouette in the center of the QTTD.

2-19. RANGE FIRING COURSES

Range firing is conducted after the firers have satisfactorily completed preparatory marksmanship training. The range firing courses are:

a. **Instructional**. Instructional firing is practice firing on a range, using the assistance of a coach.

(1) All personnel authorized or required to fire the pistol receive 12 hours of preliminary instruction that includes the following:

- Disassembly and assembly.
- Loading, firing, unloading, and immediate action.
- Preparatory marksmanship.
- Care and cleaning.

(2) The tables fired for instructional practice are prescribed in the combat pistol qualification course in Appendix A. During the instructional firing, the CPQC is fired with a coach or instructor.

b. **Combat Pistol Qualification**. The CPQC stresses the fundamentals of quick fire. It is the final test of a soldier's proficiency and the basis for his marksmanship classification. After the soldier completes the instructional practice firing, he shoots the CPQC for record. Appendix A provides a detailed description of the CPQC tables, standards, and conduct of fire. TC 25-8 provides a picture of the course.

NOTE: The alternate pistol qualification course (APQC) can be used for sustainment/ qualification if the CPQC is not available (see Appendix B).

c. **Military Police Firearms Qualification**. The military police firearms qualification course is a practical course of instruction for police firearms training (see FM 19-10).

Section IV. SAFETY

Safety must be observed during all marksmanship training. Listed below are the precautions for each phase of training. It is not intended to replace AR 385-63 or local range regulations. Range safety requirements vary according to the requirements of the course of fire. It is mandatory that the latest range safety directives and local range regulations be consulted to determine current safety requirements.

2-20. REQUIREMENTS

The following requirements apply to all marksmanship training.

a. Display a red flag prominently on the range during all firing.

b. Soldiers must handle weapons carefully and never point them at anyone except the enemy in actual combat.

c. Always assume a weapon is loaded until it has been thoroughly examined and found to contain no ammunition.

d. Indicate firing limits with red and white striped poles visible to all firers.

e. Never place obstructions in the muzzle of any weapon about to be fired.

f. Keep weapons in a prescribed area with proper safeguards.

g. Refrain from smoking on the range near ammunition, explosives, or flammables.

2-21. BEFORE FIRING

The following requirements must be met before conducting marksmanship training.

a. Close and post guards at all prescribed roadblocks and barriers.

b. Ensure all weapons are clear of ammunition and obstructions, and all slides are locked to the rear.

c. Brief all firers on the firing limits of the range and firing lanes. Firers must keep their fires within prescribed limits.

d. Ensure all firers receive instructions on know how to load and unload the weapon and on safety features.

e. Brief all personnel on all safety aspects of fire and of the range pertaining to the conduct of the courses.

f. No one moves forward of the firing line without permission of the tower operator, safety officer, or OIC.

g. Weapons are loaded and unlocked only on command from the tower operator except during conduct of the courses requiring automatic magazine changes.

h. Weapons are not handled except on command from the tower operator.

i. Firers must keep their weapons pointed downrange when loading, preparing to fire, or firing.

2-22. DURING FIRING

The following requirements apply during marksmanship training.

a. A firer does not move from his position until his weapon has been cleared by safety personnel and placed in its proper safety position. An exception is the assault phase.

b. During Table 5 of the CPQC, firers remain on line with other firers on their right or left.

c. Firers must fire only in their own firing lane and must not point the weapon into an adjacent lane, mainly during the assault phase.

d. Firers treat the air-operated pistol as a loaded weapon, observing the same safety precautions as with other weapons.

e. All personnel wear helmets during live-fire exercises.

f. Firers hold the weapon in the raised position except when preparing to fire. They then hold weapons in the ready position, pointed downrange.

2-23. AFTER FIRING

Safety personnel inspect all weapons to ensure they are clear. A check is conducted to determine if any brass or live ammunition is in the possession of the soldiers. Once cleared, pistols are secured with the slides locked to the rear.

2-24. INSTRUCTIONAL PRACTICE AND RECORD QUALIFICATION FIRING

During these phases of firing, safety personnel ensure that--

a. The firer understands the conduct of the exercise.

b. The firer has the required ammunition and understands the commands for loading and unloading.

c. The firer complies with all commands from the tower operator.

d. Firers maintain proper alignment with other firers while moving downrange.

e. Weapons are always pointed downrange.

f. Firers fire within the prescribed range limits.

g. Weapons are cleared after each phase of firing, and the tower-operator is aware of the clearance.

h. Malfunctions or failures to fire that are due to no fault of the firer are reported immediately. On command of the tower operator, the weapon is cleared and action is taken to allow the firer to continue with the exercise.

NOTE: For training and qualification standards, see Appendixes A through D.

APPENDIX A
COMBAT PISTOL QUALIFICATION COURSE

**This appendix explains the combat pistol qualification course (CPQC). If it is unavailable, the alternate pistol qualification course (APQC) may be used to sustain training and to qualify firers.*

The tower operator is completely responsible for and in charge of the range and the course. He controls absolutely all activities related to firing. The tower operator tells the scorers what to do when, for example, when to issue the preloaded magazines to firers. Only the tower operator may issue the order to fire. Scorers and firers must await the tower operator's orders.

A-1. COURSE INFORMATION

The CPQC (shown in TC 25-8) requires the Soldier to engage single and multiple targets at various ranges using the fundamentals of quick fire.

a. **Extra Rounds**. For each table of the CPQC, the firer is given extra rounds to reengage missed targets. Although only 30 targets will be exposed during the entire course, each firer will receive 40 rounds of ammunition. Hitting a target with an additional round during the exposure time is just as effective as hitting it with the first round. Consequently, the firer is not penalized for using or not using the extra ammunition. However, any unused ammunition must be turned in at the end of the table, and may not be used in any other table.

b. **Magazine Changes**. Only three magazine changes are required during this course: one change in Firing Table II, and two changes in Firing Table V. For safety, each of these two tables begins with a magazine loaded only with 1 round. The first target appears, and the firer engages it with that round. By the time another target appears 8 seconds later, the firer must have reloaded and prepared to engage. He will receive no commands to reload. Failure to reload in time to engage the second target is scored as a miss. This teaches the Soldier to change magazines instinctively, quickly, and safely under pressure. In Table V, a second magazine change is commanded by the control tower.

c. **Double-Action Mode**. When firing the 9-mm pistol, the Soldier uses double-action to fire the first round in every table.

d. **Range to Target**. The range to exposed targets must not exceed 31 meters from the firer. Table A-1 shows target exposure times for each firing table.

<table>
<tr><th rowspan="2">NUMBER OF TARGETS</th><th colspan="7">FIRING TABLE</th></tr>
<tr><th>I</th><th>II</th><th>III</th><th>IV</th><th>V</th><th>VI</th><th>VII</th></tr>
<tr><td>Single targets</td><td colspan="3">3 Sec</td><td colspan="2">2 Sec</td><td colspan="2">10 Sec</td></tr>
<tr><td>Multiple targets</td><td colspan="3">5 Sec</td><td colspan="2">4 Sec</td><td colspan="2">20 Sec</td></tr>
</table>

Table A-1. Target-exposure times.

A-2. STANDARDS BY FIRING TABLE

The following qualification tables apply for day, night, and CBRN qualification. The standing firing position is used throughout the qualification:

NOTE:
1. The range OIC determines a common target sequence for all lanes. This keeps a firer from getting ahead of adjacent firers.
2. Target sequences vary in distance from the firer, starting with no more than two targets at 10 meters and the farthest targets at 31 meters.
3. The firer will remain in the same firing lane throughout the CPQC.

a. **Table I--Day Standing**. For this table, the firer receives one magazine with 7 rounds in it. Five targets (single) are exposed. The firer assumes the standing firing position at the firing line. He holds the weapon at the ready. The tower operator sets the target sequence.

b. **Table II--Day Standing**. For this table, the firer receives two magazines: one containing 1 round, and the other containing 7 rounds. Six targets (four single and one set of two) are exposed.

(1) ***First Magazine***. The firer loads the first magazine (containing 1 round). One target is exposed.

(2) ***Second Magazine***. After he fires the round in the first magazine, the firer must change magazines at once. He has 8 seconds to load the second magazine (containing 7 rounds) and prepare to fire before the next target is exposed. Once it appears, he must engage in the 3 seconds before it is lowered. Failure to do so is scored as a miss.

c. **Table III--Day Standing**. For this table, the firer receives one magazine containing 7 rounds. Five targets (three single and one set of two) are exposed.

d. **Table IV--Day Standing**. For this table, the firer receives one magazine containing 5 rounds. Four targets (two single and one set of two) are exposed.

e. **Table V--Day Moving Out**. For this table, the firer receives three magazines: one each with one, seven, and 5 rounds. Ten targets are exposed. The firer begins 10 meters behind the firing line, in the middle of the trail.

(1) The firer loads the first magazine (containing 1 round). He places the second magazine (containing 7 rounds) in the magazine pouch closest to his firing hand. He places third magazine (containing 5 rounds) in the magazine pouch farthest from his firing hand.

(2) When the firer reaches the firing line, a single target is exposed. The firer has 2 seconds to hit it before it is lowered. He then has 8 seconds to load the second magazine (containing 7 rounds).

(3) At the end of 8 seconds, another single target is exposed to the firer. If the firer has not loaded the second magazine in time to engage this target, this round is scored as a miss.

(4) When the tower operator is sure that the firing line has completed the magazine change, he commands MOVE OUT. He then exposes two multiple targets, one after the other, at various ranges from the firer.

(5) After two sets of multiple targets are exposed, the Soldier is commanded to load the 5-round magazine. After the command MOVE OUT is given, the remaining targets are presented to the firer in sequence. After the last targets are hit or lowered, the firer clears the weapon.

(6) The firer holds the weapon in the raised pistol position with the slide to the rear. He returns to the starting point and places the weapon on the stand. He turns in any excess

ammunition to the ammunition point. On hearing the order to do so, he moves to the firing line.

f. **Table VI--Day Standing, CBRN**. All firers will wear protective masks with hoods. For this table, the firer receives one magazine containing 7 rounds. Five targets (three single and one set of two) are exposed.

g. **Table VII--Night Standing**. For this table, the firer receives one magazine containing 5 rounds. Four targets (two single and one set of two) are exposed.

NOTE: Commanders may use the Engagement Skills Trainer (EST) 2000 to conduct Firing Tables VI and VII (CBRN and night fire).

A-3. TOWER OPERATOR'S AUTHORITY

The tower operator is responsible for the range. For this reason, only he can give orders to scorers and firers on the range.

A-4. CONDUCT OF FIRE BY FIRING TABLE

For each table, the tower operator has scorers issue only the rounds required for that table. The following fire commands show how the tower operator runs range fire on the CPQC:

a. **Table I--Day Standing**. The tower operator orders firers to move to the firing line in preparation for firing. He orders the firers to position themselves next to the weapon stands and secure their weapons. On command, the scorer issues to the firer one magazine containing 7 rounds.

(1) The tower operator commands--

TABLE ONE, STANDING POSITION, 7 ROUNDS.
LOAD AND LOCK.
READY ON THE RIGHT.
READY ON THE LEFT.
READY ON THE FIRING LINE.
UNLOCK YOUR WEAPONS.
FIRERS, WATCH YOUR LANE.

(2) The tower operator exposes the targets to the firers. When all targets have been exposed and engaged or lowered, the tower operator commands--

CEASE FIRE.
ARE THERE ANY ALIBIS? (ALIBIS GET 10 SECONDS FOR EACH ROUND NOT FIRED.)
CLEAR ALL WEAPONS.
CLEAR ON THE RIGHT.
CLEAR ON THE LEFT.
THE FIRING LINE IS CLEAR.
FIRERS, PLACE YOUR WEAPONS ON THE STANDS WITH SLIDES LOCKED TO THE REAR.
FIRERS AND SCORERS, MOVE DOWNRANGE AND CHECK YOUR TARGETS.
MARK AND COVER ALL HOLES.

b. **Table II--Day Standing**. The tower operator orders firers to secure their weapons. On command, the scorer issues to the firer one magazine containing a single round and another magazine containing 7 rounds.

(1) The tower operator commands--

TABLE TWO, STANDING POSITION, EIGHT ROUNDS.
LOAD AND LOCK ONE MAGAZINE WITH 1 ROUND.
LOAD YOUR 7-ROUND MAGAZINE WITHOUT COMMAND.
READY ON THE RIGHT.
READY ON THE LEFT.
READY ON THE FIRING LINE.
UNLOCK YOUR WEAPONS.
FIRERS, WATCH YOUR LANES.

(2) The tower operator exposes the targets to the firers. When all targets have been exposed and engaged or lowered, the tower operator commands--

CEASE FIRE.
ARE THERE ANY ALIBIS? (ALIBIS GET 10 SECONDS FOR EACH ROUND NOT FIRED.)
CLEAR ALL WEAPONS.
CLEAR ON THE RIGHT.
CLEAR ON THE LEFT.
THE FIRING LINE IS CLEAR.
FIRERS, PLACE YOUR WEAPONS ON THE STANDS WITH SLIDES LOCKED TO THE REAR.
FIRERS AND SCORERS, MOVE DOWNRANGE AND CHECK YOUR TARGETS.
MARK AND COVER ALL HOLES.

c. **Table III--Day Standing**. The tower operator orders the firers to secure their weapons. On command, the scorer issues to the firer one magazine containing 7 rounds.

(1) The tower operator commands--

TABLE THREE, STANDING POSITION, 7 ROUNDS.
LOAD AND LOCK.
READY ON THE RIGHT.
READY ON THE LEFT.
READY ON THE FIRING LINE.
UNLOCK YOUR WEAPONS.
FIRERS, WATCH YOUR LANES.

(2) The tower operator exposes the targets to the firers. When all targets have been exposed and engaged or lowered, the tower operator commands--

CEASE FIRE.
ARE THERE ANY ALIBIS? (ALIBIS GET 10 SECONDS FOR EACH ROUND NOT FIRED.)
CLEAR ALL WEAPONS.
CLEAR ON THE RIGHT.
CLEAR ON THE LEFT.
THE FIRING LINE IS CLEAR.
FIRERS, PLACE YOUR WEAPONS ON THE STANDS WITH SLIDES LOCKED TO THE REAR.
FIRERS AND SCORERS, MOVE DOWNRANGE AND CHECK YOUR TARGETS.
MARK AND COVER ALL HOLES.

d. **Table IV--Day Standing**. The tower operator orders the firers to secure their weapons. On command, the scorer issues to the firer one magazine containing 5 rounds.

(1) The tower operator commands--

TABLE FOUR, STANDING POSITION, 5 ROUNDS.
LOAD AND LOCK.
READY ON THE RIGHT.
READY ON THE LEFT.
READY ON THE FIRING LINE.
UNLOCK YOUR WEAPONS.
FIRERS, WATCH YOUR LANES.

(2) The tower operator exposes the targets to the firers. When all targets have been exposed and engaged or lowered, the tower operator commands--

CEASE FIRE.
ARE THERE ANY ALIBIS? (ALIBIS GET 10 SECONDS FOR EACH ROUND NOT FIRED)
CLEAR ALL WEAPONS.
CLEAR ON THE RIGHT.
CLEAR ON THE LEFT.
THE FIRING LINE IS CLEAR.
FIRERS, PLACE YOUR WEAPONS ON THE STAND WITH SLIDES LOCKED TO THE REAR.
FIRERS AND SCORERS, MOVE DOWNRANGE AND CHECK YOUR TARGETS.
MARK AND COVER ALL HOLES.

e. **Table V--Day Moving Out**. The tower operator orders the firers to secure their weapons and move to the center of the trail 10 meters behind the firing line. On command, the scorer issues to the firer one magazine containing 1 round; a second magazine containing 7 rounds; and a third magazine containing 5 rounds.

(1) The tower operator commands--

TABLE FIVE, STANDING POSITION, THIRTEEN ROUNDS.
LOAD AND LOCK ONE MAGAZINE WITH 1 ROUND.
LOAD YOUR SEVEN AND 5-ROUND MAGAZINES AT MY COMMAND.
READY ON THE RIGHT.
READY ON THE LEFT.
READY ON THE FIRING LINE.
PISTOLS AT THE READY POSITION.
UNLOCK YOUR WEAPON.
FIRERS, WATCH YOUR LANES.
MOVE OUT.

(2) The tower operator exposes the targets to the firers. After each target or group of targets has been engaged, he commands--

WEAPONS AT THE READY POSITION.
MOVE OUT.

(3) After the firers complete Table V, the tower operator commands--

CEASE FIRE.
ARE THERE ANY ALIBIS? (ALIBIS GET 10 SECONDS FOR EACH ROUND NOT FIRED)
CLEAR ALL WEAPONS.
CLEAR ON THE RIGHT.
CLEAR ON THE LEFT.
THE FIRING LINE IS CLEAR.
FIRERS, KEEP YOUR WEAPONS UP AND POINTED DOWNRANGE.
SCORERS AND FIRERS, MOVE BACK TO THE FIRING LINE, AND PLACE YOUR WEAPONS ON THE STAND WITH SLIDES LOCKED TO THE REAR..
FIRERS AND SCORERS, MOVE DOWNRANGE AND CHECK YOUR TARGETS.
MARK AND COVER ALL HOLES.

(3) The tower operator has each scorer total the firer's scorecard and turn it in to the range officer or his representative. The firing orders are rotated and the above sequence continued until all orders have fired.

f. **Table VI--Day Standing, CBRN**. The firer will wear a protective mask with hood.

(1) The tower operator orders the firers to position themselves next to the weapon stands. On command, the scorer issues to the firer one magazine containing 7 rounds. The firer must get three hits to receive a "GO" on this table. The tower operator commands--

TABLE SIX, CBRN FIRE, STANDING POSITION, 7 ROUNDS.
LOAD AND LOCK.
READY ON THE RIGHT.
READY ON THE LEFT.
READY ON THE FIRING LINE.
UNLOCK YOUR WEAPONS.
FIRERS, WATCH YOUR LANES.

(2) The tower operator exposes the targets to the firers. When all targets have been exposed and engaged or lowered, the tower operator commands--

CEASE FIRE.
ARE THERE ANY ALIBIS? (ALIBIS GET 10 SECONDS FOR EACH ROUND NOT FIRED)
CLEAR ALL WEAPONS.
CLEAR ON THE RIGHT.
CLEAR ON THE LEFT.
THE FIRING LINE IS CLEAR.
FIRERS, PLACE YOUR WEAPONS ON THE STAND, WITH SLIDES LOCKED TO THE REAR..
FIRERS AND SCORERS, MOVE DOWNRANGE AND CHECK YOUR TARGETS.
MARK AND COVER ALL HOLES.

g. **Table VII--Night Standing**. The tower operator orders the firers to position themselves next to the weapon stands. On command, the scorer issues to the firer one magazine containing 5 rounds. The firer must get two hits to receive a "GO" on this table.

(1) The tower operator commands--

TABLE SEVEN, NIGHT FIRE, STANDING POSITION, 5 ROUNDS.
LOAD AND LOCK.
READY ON THE RIGHT.
READY ON THE LEFT.
READY ON THE FIRING LINE.
UNLOCK YOUR WEAPONS.
FIRERS, WATCH YOUR LANES.

(2) The tower operator exposes the targets to the firers. When all targets have been exposed and engaged or lowered, The tower operator commands--

CEASE FIRE.
ARE THERE ANY ALIBIS? (ALIBIS GET 10 SECONDS FOR EACH ROUND NOT FIRED)
CLEAR ALL WEAPONS.
CLEAR ON THE RIGHT.
CLEAR ON THE LEFT.
THE FIRING LINE IS CLEAR.
FIRERS, PLACE YOUR WEAPONS ON THE STAND, WITH SLIDES LOCKED TO THE REAR.
FIRERS AND SCORERS, MOVE DOWNRANGE AND CHECK YOUR TARGETS.
MARK AND COVER ALL HOLES.

A-5. ALIBIS

Alibis are fired after each table exactly where they occurred. Firers get 10 seconds for each alibi. The same fire commands apply to alibis. If a weapon or target malfunctions while the firer is firing from a stationary position, he reports the malfunction. He keeps his weapon pointed up and downrange. Should the malfunction occur during Table V, the firer keeps his weapon pointed up and downrange, but he continues to move forward, keeping himself aligned with the firers to his right and left.

A-6. RULES

Certain rules apply to the conduct of fire during the CPQC:

a. **Assistance**. During instructional fire, the coach and assistant instructors should assist the firer in correcting errors. However, during record fire, no one may help or try to help the firer while or after he takes his position at the firing point.

b. **Accidental Discharges**. After the firer takes his place on the firing lane, every shot counts. Even if he fires away from the target or discharges the weapon accidentally, then that counts as his shot. He receives no replacement round or second chance.

c. **Fire on the Wrong Target**. Each firer observes the location of the target in his own lane. Shots fired on the wrong target count as a miss. A firer is credited only for the targets he hits in his own firing lane.

d. **Fire After the Signal to Lower Targets**. Any shot after the target starts to lower is scored as a miss.

e. **Extra Shot Fired at an E-Type Silhouette Target**. If the firer hits the target while the target is exposed, that is, before it begins to lower, then he receives credit for the hit. The number of rounds fired to obtain the hit does not matter.

f. **Excess Ammunition**. At the end of each firing table, the firer turns in any excess ammunition. This ammunition is not re-issued to him for use in the other firing tables.

g. **Target Sequence**. The tower operator sets a common target sequence for all lanes. This keeps a firer from getting ahead of the firers in adjacent lanes. Target sequence varies in distance from the firer. It starts with 31 meters and allows for no more than two 7-meter targets.

A-7. SCORECARD

Figure A-1 shows an example completed DA Form 88-R (*Combat Pistol Qualification Course Scorecard*), and a blank copy is provided in the back of the book. The blank form may be reproduced locally on 8 1/2- by 11-inch paper. It may also be downloaded from the Internet at Army Knowledge Online (http://www.army.mil/usapa/eforms/). The scorecard lists the standards and provides scoring grids for the CPQC.

COMBAT PISTOL QUALIFICATION COURSE SCORECARD

For use of this form, see FM 3-23.35; the proponent agency is TRADOC.

1. ID CODE [1]	2. UNIT	3. LANE NO.	4. ORDER	5. GROUP	6. DATE *(YYYYMMDD)*
20070426LUS01	C CO 2/29 IN	3	4	5	20070426

EXAMPLE

7. TABLE I — DAY STANDING [2] — 1 Mag--7 Rds

TIME	TGT	HITS
3 Sec	1	X
3 Sec	2	X
3 Sec	3	X
3 Sec	4	X
3 Sec	5	
Table I Total Hits		4

8. TABLE II — DAY STANDING [2] — Mag 1--1 Rd [3], Mag 2--7 Rds

TIME	TGT	HITS
3 Sec	1	X
3 Sec	2	X
5 Sec	3	X
	4	X
3 Sec	5	
3 Sec	6	X
Table II Total Hits		5

9. TABLE III — DAY STANDING [2] — 1 Mag--7 Rds

TIME	TGT	HITS
3 Sec	1	X
3 Sec	2	X
3 Sec	3	X
5 Sec	4	X
	5	X
Table III Total Hits		5

10. TABLE IV — DAY STANDING [2] — 1 Mag--5 Rds

TIME	TGT	HITS
2 Sec	1	X
2 Sec	2	
4 Sec	3	X
	4	X
Table IV Total Hits		3

11. TABLE V — DAY MOVING OUT [2] — Mag 1--1 Rd [3], Mag 2--7 Rds, Mag 3--5 Rds [4]

TIME	TGT	HITS
2 Sec	1	X
2 Sec	2	X
4 Sec	3	X
	4	
4 Sec	5	X
	6	X
2 Sec	7	X
2 Sec	8	X
4 Sec	9	X
	10	X
Table V Total Hits		9

12. TABLE VI — DAY STANDING CBRN [2] — 1 Mag--7 Rds

TIME	TGT	GO	NO-GO
10 Sec	1		X
10 Sec	2	X	
10 Sec	3	X	
20 Sec	4	X	
	5	X	
GO or NO-GO [5]		X	

13. TABLE VII — NIGHT STANDING [2] — 1 Mag--5 Rds

TIME	TGT	GO	NO-GO
10 Sec	1	X	
10 Sec	2	X	
20 Sec	3	X	
	4	X	
GO or NO-GO [5]		X	

14. NOTES

[1] Do not use personal information.

[2] Firer receives the number of rounds required to fire a specific table. The OIC of firing sets precedures for loading and unloading.

[3] Firer has 8 seconds to change the magazine.

[4] Controlled change.

[5] Check GO or NO-GO for whole table.

- Three hits are required for a GO on Table VI.
- Two hits are required for a GO on Table VII.

15. RATING CALCULATOR

Enter total hits from Tables I through V and add them to obtain a grand total:

TABLE I	4	
TABLE II	5	
TABLE III	5	
TABLE IV	3	
TABLE V	9	+
GRAND TOTAL	26	=

16. RATING SCALE

GRAND TOTAL *(BLOCK 15)*	RATING	
26 - 30	EXPERT	X
21 - 25	FIRST CLASS	
16 - 20	SECOND CLASS	
0 - 15	UNQUALIFIED	

17. GRADER'S INITIALS	18. DATE *(YYYYMMDD)*	19. OIC'S INITIALS	20. DATE *(YYYYMMDD)*
AMR	20070426	RVL	20070426

DA FORM 88-R, MAR 2007 PREVIOUS EDITIONS ARE OBSOLETE. APD V1.00

Figure A-1. Example completed DA Form 88-R, *Combat Pistol Qualification Course Scorecard.*

NOTE: Numbers in the "**TGT**" columns do not represent a particular sequence in which the targets will appear. They just identify how *many* targets the firer will engage in each firing table.

a. Each time a firer hits or kills a target, the scorer places an "X" in the "**HITS**" column and an "M" for miss. Each hit is worth 1 point. After the firer finishes firing, the scorer totals and signs the scorecard.

b. The following qualification standards are shown in Block 16 (RATING SCALE) on the scorecard:

- Expert--26 to 30 hits.
- Sharpshooter--21 to 25 hits.
- Marksman--16 to 20 hits.
- Unqualified--0 to 15 hits.

c. The CBRN and night firing tables are scored as GO or NO-GO. The firer either qualifies on those tables, or not. For each of these tables (VI and VII), the firer gets a GO if he hits the target, and a NO-GO if he misses.

d. To qualify, the firer must earn a minimum total score of 16 on Tables I through V, three hits on Table VI (CBRN), and two hits on Table VII (Night).

A-8. TARGETS

Each firing lane requires seven electrical, device-type targets as well as a single E-type silhouette. Aggressor figures may be superimposed on the silhouettes to add realism to the course of fire.

A-9. QUICK-FIRE TARGET TRAINING DEVICE

The unit can get a quick-fire target-training device (QTTD) locally. To ensure standardization, quality, durability, and appearance, the device should be constructed by a qualified organization with documented experience producing similar devices such as the training aids section of the local Training Support Center.

APPENDIX B

ALTERNATE PISTOL QUALIFICATION COURSE

**Once the soldier completes instructional fire, he must complete the combat pistol qualification course for the record. However, when the CPQC is unavailable, the alternate pistol qualification course may be used.*

The tower operator is completely responsible for and in charge of the range and the course. He controls all *activities related to firing. The tower operator tells the scorers what to do, when. For example, he tells them when to issue the preloaded magazines to firers.*

Only *the tower operator may issue the order to fire. Scorers and firers* must *await the tower operator's orders.*

B-1. CONDITIONS AND STANDARDS

The firer is given 40 rounds of ammunition for Tables I through IV, and 14 rounds for Tables V and VI:

a. **Table I--Day Standing**. Given one 7-round magazine in daylight, within 21 seconds, engage the 25-meter APQC target from the standing position with 7 rounds of ammunition.

b. **Table II--Day Kneeling**. Given one 6-round magazine and one 7-round magazine in daylight, within 45 seconds, engage the 25-meter APQC target from the kneeling position with all 13 rounds.

(1) From a standing position, assume a good kneeling position and engage the target with all 6 rounds in the first magazine.

(2) Perform a rapid magazine change.

(3) Engage the target with all 7 rounds in the second magazine.

c. **Table III--Day Crouching**. Given two magazines with 5 rounds each in daylight, within 35 seconds, engage the 25-meter APQC target from the crouching position with all 10 rounds.

(1) From a standing position, assume a good crouching position, and engage the target with one 5-round magazine.

(2) Perform a rapid magazine change.

(3) Engage the target with the second 5-round magazine.

d. **Table IV--Day Prone**. Given two magazines with 5 rounds each in daylight, within 35 seconds, engage the 25-meter APQC target from the prone position with all 10 rounds.

(1) From a standing position, assume a good prone position, and engage the target with one 5-round magazine

(2) Perform a rapid magazine change.

(3) Engage the target with the second 5-round magazine.

e. **Table V--Day CBRN Crouching**. Given one 7-round magazine in simulated CBRN conditions, within 21 seconds, engage the 25-meter target from the crouching position with all 7 rounds.-

f. **Table VI--Night Crouching.** Given one 5-round magazine under night conditions, within 21 seconds, engage a 25-meter target from a crouching position with all 5 rounds.

B-2. CONDUCT OF FIRE

The fire commands shape the conduct of range fire in the APQC. When the firer is using a 9-mm pistol, he fires the first round in each table in double-action mode. At the end of each firing table, each firer turns in any excess ammunition to the scorer. He may not use it in other firing tables. At the end of the course, the scorer turns in all excess ammunition to the ammunition point.

NOTES: 1. Commanders may use the Engagement Skills Trainer (EST) 2000 for Tables V *(Day CBRN Crouching)* and VI (*Night Crouching*).

2. Only the tower operator may give firing instructions.

a. **Table I--Day Standing**. The tower operator gives the order to move to the firing line and to prepare to fire.

(1) On the tower operator's command, the scorer issues to the firer one magazine containing 7 rounds.

(2) The tower operator then commands--

TABLE ONE, STANDING POSITION, 7 ROUNDS.
LOAD AND LOCK ONE SEVEN ROUND MAGAZINE.
IS THE FIRING LINE READY? (FIRERS USING 9-MM PISTOLS PLACE THEM ON DOUBLE ACTION.)
READY ON THE RIGHT.
READY ON THE LEFT.
THE FIRING LINE IS READY.
FIRERS, UNLOCK YOUR WEAPONS.
FIRERS, WATCH YOUR LANES.

(3) At the end of the prescribed firing time, the tower operator commands--

CEASE FIRE.
ARE THERE ANY ALIBIS? (ALIBIS GET 8 SECONDS FOR EACH ROUND NOT FIRED.)
UNLOAD AND CLEAR ALL WEAPONS.
IS THE FIRING LINE CLEAR?
CLEAR ON THE RIGHT.
CLEAR ON THE LEFT.
THE FIRING LINE IS NOW CLEAR.
FIRERS, PLACE YOUR WEAPONS ON THE STANDS WITH SLIDES LOCKED TO THE REAR.
FIRERS AND SCORERS, MOVE DOWNRANGE AND CHECK YOUR TARGETS.
MARK AND COVER ALL HOLES.

b. **Table II--Day Kneeling**. The tower operator orders firers to move up to the firing line. On the tower operator's command, the scorer issues two magazines, one loaded with 6 rounds and the other with 7 rounds, to the firer.

(1) The tower operator then commands--

TABLE TWO, KNEELING POSITION, 6 ROUNDS.
LOAD AND LOCK ONE 6-ROUND MAGAZINE.
LOAD YOUR 7-ROUND MAGAZINES WITHOUT COMMAND
IS THE FIRING LINE READY? (FIRERS USING 9-MM PISTOLS PLACE THEM ON DOUBLE- ACTION.)
READY ON THE LEFT.
READY ON THE RIGHT.
THE FIRING LINE IS READY.
FIRERS, UNLOCK YOUR WEAPONS.
FIRERS, WATCH YOUR LANES.

(2) At the end of the prescribed firing time, the tower operator commands--

CEASE FIRE.
ARE THERE ANY ALIBIS? (ALIBIS GET 8 SECONDS FOR EACH ROUND NOT FIRED.)
UNLOAD AND CLEAR ALL WEAPONS.
IS THE FIRING LINE CLEAR?
CLEAR ON THE RIGHT.
CLEAR ON THE LEFT.
THE FIRING LINE IS NOW CLEAR.
FIRERS, PLACE YOUR WEAPONS ON THE STANDS WITH SLIDES LOCKED TO THE REAR.
FIRERS AND SCORERS, MOVE DOWNRANGE AND CHECK YOUR TARGETS.
MARK AND COVER ALL HOLES.

c. **Table III--Day Crouching**. The tower operator orders firers to move to the firing line.

(1) On the tower operator's command, the scorer issues two 5-round magazines to the firer:

TABLE THREE, CROUCHING POSITION, 5 ROUNDS.
LOAD AND LOCK ONE 5-ROUND MAGAZINE.
LOAD YOUR SECOND 5-ROUND MAGAZINE WITHOUT COMMAND.
IS THE FIRING LINE READY? (FIRERS USING 9-MM PISTOLS PLACE THEM ON DOUBLE- ACTION.)
READY ON THE RIGHT.
READY ON THE LEFT.
THE FIRING LINE IS READY.
FIRERS, WATCH YOUR LANES.

(2) At the end of the prescribed firing time, the tower operator commands—

CEASE FIRE.
ARE THERE ANY ALIBIS? (ALIBIS GET 10 SECONDS FOR EACH ROUND NOT FIRED.)
UNLOAD AND CLEAR ALL WEAPONS.
IS THE FIRING LINE CLEAR?
CLEAR ON THE RIGHT.
CLEAR ON THE LEFT.
THE FIRING LINE IS NOW CLEAR.
FIRERS, PLACE YOUR WEAPONS ON THE STANDS WITH SLIDES LOCKED TO THE REAR.
FIRERS AND SCORERS, MOVE DOWNRANGE AND CHECK YOUR TARGETS.
MARK AND COVER ALL HOLES.

d. **Table IV--Day Prone Unsupported**. The tower operator orders firers to move to the firing line.

(1) On the tower operator's command, the scorer issues two 5-round magazines to the firer. After the firer completes Table IV, the scorer and firer repair or replace targets for the next firing order:

TABLE FOUR, PRONE UNSUPPORTED POSITION, 5 ROUNDS.
LOAD AND LOCK ONE 5-ROUND MAGAZINE.
LOAD YOUR SECOND 5-ROUND MAGAZINE WITHOUT COMMAND.
IS THE FIRING LINE READY?
READY ON THE RIGHT.
READY ON TH E LEFT.
THE FIRING LINE IS READY.
FIRERS, UNLOCK YOUR WEAPONS.
FIRERS, WATCH YOUR LANE.

(2) At the end of the prescribed firing time, the tower operator commands--

CEASE FIRE.
ARE THERE ANY ALIBIS? (ALIBIS GET 10 SECONDS FOR EACH ROUND NOT FIRED.)
UNLOAD AND CLEAR ALL WEAPONS.
IS THE FIRING LINE CLEAR?
CLEAR ON THE RIGHT.
CLEAR ON THE LEFT.
THE FIRING LINE IS NOW CLEAR.
FIRERS, PLACE YOUR WEAPONS ON THE STANDS WITH SLIDES LOCKED TO THE REAR.
FIRERS AND SCORERS, MOVE DOWNRANGE AND CHECK YOUR TARGETS.
MARK AND COVER ALL HOLES.

e. **Table V--Day CBRN Crouching.** All firers wear protective masks with hoods. The tower operator orders firers to move to the firing line. On the tower operator's command, the scorer issues one 7-round magazine to the firer.

(1) He issues the same commands he did for Table I.

TABLE FIVE, CBRN FIRE, CROUCHING POSITION, 7 ROUNDS.
LOAD AND LOCK ONE MAGAZINE.
IS THE FIRING LINE READY? (FIRERS USING 9-MM PISTOLS PLACE THEM ON DOUBLE- ACTION.)
READY ON THE RIGHT.
READY ON THE LEFT.
THE FIRING LINE IS READY.
FIRERS, WATCH YOUR LANE.

(2) At the end of the prescribed firing time, the tower operator commands--

CEASE FIRE.
ARE THERE ANY ALIBIS? (ALIBIS GET 10 SECONDS FOR EACH ROUND NOT FIRED.)
UNLOAD AND CLEAR ALL WEAPONS.
IS THE FIRING LINE CLEAR?
CLEAR ON THE RIGHT.
CLEAR ON THE LEFT.
THE FIRING LINE IS NOW CLEAR.
FIRERS, PLACE YOUR WEAPONS ON THE STANDS WITH SLIDES LOCKED TO THE REAR.
FIRERS AND SCORERS, MOVE DOWNRANGE AND CHECK YOUR TARGETS.
MARK AND COVER ALL HOLES.

*f. **Table VI--Night Crouching.** The tower operator orders firers to move to the firing line and to prepare to fire. On the tower operator's command, the scorer issues one 5-round magazine to the firer.

(1) The tower operator commands--

TABLE SIX, NIGHT FIRE, CROUCHING POSITION, 5 ROUNDS.
LOAD AND LOCK ONE MAGAZINE.
IS THE FIRING LINE READY? (FIRERS USING 9-MM PISTOLS PLACE THEM ON DOUBLE -ACTION.)
READY ON THE RIGHT.
READY ON THE LEFT.
THE FIRING LINE IS READY.
FIRERS, UNLOCK YOUR WEAPONS.
FIRERS, WATCH YOUR LANE.

(2) At the end of the prescribed firing time, the tower operator commands--

CEASE FIRE.
ARE THERE ANY ALIBIS? (ALIBIS GET 8 SECONDS FOR EACH ROUND NOT FIRED.)
UNLOAD AND CLEAR ALL WEAPONS.
IS THE FIRING LINE CLEAR?
CLEAR ON THE RIGHT.
CLEAR ON THE LEFT.
THE FIRING LINE IS NOW CLEAR.
FIRERS, PLACE YOUR WEAPONS ON THE STANDS WITH SLIDES LOCKED TO THE REAR.
FIRERS AND SCORERS, MOVE DOWNRANGE AND CHECK YOUR TARGETS.
MARK AND COVER ALL HOLES.

B-3. ALIBIS

The scorer reports and records any weapon or target malfunction that occurs during fire. The firer is allowed one alibi at the end of each table. For Tables I through IV, he is allowed 8 seconds for each alibi. For Tables V and VI, he is allowed 10 seconds each. All alibis are fired where the malfunction occurred, using the same firing commands.

B-4. SCORECARD

Figure B-1 shows a 25–meter, E-type silhouette. Figure B-2 shows an example completed DA Form 5704-R (*Alternate Pistol Qualification Course Scorecard*). A blank copy is provided in the back of the book. The blank form may be reproduced locally on 8 1/2 by 11-inch paper. It may also be downloaded from http://www.army.mil/usapa/eforms/.

a. **Firing Tables I thru IV**. Each time a firer hits or kills a target, the scorer places an "X" (hit) or "M" (miss) in the appropriate box. Then, he writes the total number of hits and misses at the bottom of each column. After the firer completes the first four firing tables, the scorer enters the totals from each of the four tables in Block 14 (RATING CALCULATOR), and adds them to get the grand total of all hits. He marks the appropriate rating in Block 15 (RATING SCALE).

*EXPERT--36 to 40 hits.
*SHARPSHOOTER--29 to 35 hits.
*MARKSMAN--24 to 28 hits.
*UNQUALIFIED--0 to 23 hits.

b. **Firing Tables V and VI**. Each time a firer hits or kills a target, the scorer again places an "X" or "M" in the appropriate box. Then, he writes the total number of hits and misses at the bottom of each column. The firer must hit four targets in each table to receive a "GO" on that table.

B-5. ASSISTANCE

During instructional fire, the coach and assistant instructors should help the firer correct errors. However, during record fire, no one may help the firer while or after he takes his position at the firing point.

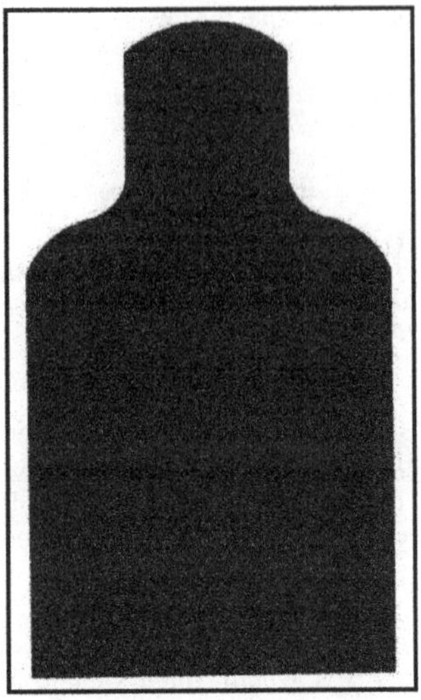

Figure B-1. E-type silhouette, 25-meter, without rings.

ALTERNATE PISTOL QUALIFICATION COURSE SCORECARD

For use of this form, see FM 3-23.35; the proponent agency is TRADOC.

1. ID CODE [1]	2. UNIT	3. LANE	4. ORDER	5. GROUP	6. DATE *(YYYYMMDD)*
20080426LUS01	C CO 2/29TH IN	3	4	5	20080426

7. TABLE I — DAY STANDING [2] — 1 Mag : 7 Rds

TIME	TGT	HITS
21 Sec	1	X
	2	X
	3	X
	4	M
	5	X
	6	X
	7	M
Table I Total Hits		

8. TABLE II — DAY KNEELING [2] — Mag 1 : 6 Rds [3] — Mag 2 : 7 Rds

TIME	TGT	HITS
45 Sec	1	X
	2	X
	3	X
	4	M
	5	X
	6	X
	7	X
	8	X
	9	X
	10	X
	11	X
	12	X
	13	X
Table II Total Hits		

9. TABLE III — DAY CROUCHING [2] — Mag 1 : 5 Rds [3] — Mag 2 : 5 Rds

TIME	TGT	HITS
35 Sec	1	X
	2	X
	3	X
	4	X
	5	X
	6	X
	7	X
	8	M
	9	X
	10	X
Table III Total Hits		

10. TABLE IV — DAY PRONE UNSUPPORTED [2] — Mag 1 : 5 Rds [3] — Mag 2 : 5 Rds

TIME	TGT	HITS
35 Sec	1	X
	2	X
	3	X
	4	X
	5	X
	6	X
	7	X
	8	X
	9	X
	10	X
Table IV Total Hits		

11. TABLE V — DAY CBRN CROUCHING [2] — 1 Mag : 7 Rds

TIME	TGT	GO	NO-GO
21 Sec	1	X	
	2		X
	3	X	
	4	X	
	5	X	
	6	X	
	7	X	
Table V Total Hits		0	0
GO or NO-GO [4]		☐	☐

12. TABLE VI — NIGHT CROUCHING [2] — 1 Mag : 5 Rds

TIME	TGT	GO	NO-GO
21 Sec	1		X
	2	X	
	3	X	
	4	X	
	5	X	
Table VI Total Hits		0	0
GO or NO-GO [4]		☐	☐

SAMPLE

13. NOTES

1 Do not use personal information.

2 Firer receives the number of rounds required to fire a specific table. The OIC of firing sets procedures for loading and unloading.

3 Firer must change magazine rapidly.

4 Check GO or NO-GO box for whole table.
- . Four hits are required for a GO on Table V.
- . Two hits are required for a GO on Table VI.

14. RATING CALCULATOR

Enter total hits from Tables I through IV and add them to obtain a grand total:

TABLE I	
TABLE II	
TABLE III	
TABLE IV	+
GRAND TOTAL	=

15. RATING SCALE

GRAND TOTAL *(BLOCK 14)*	RATING	
36 - 40	EXPERT	☐
29 - 35	FIRST CLASS	☐
24 - 28	SECOND CLASS	☐
0 - 23	UNQUALIFIED	☐

16. GRADER'S INITIALS	17. DATE *(YYYYMMDD)*	18. OIC'S INITIALS	19. DATE *(YYYYMMDD)*
	20080426		20080426

DA FORM 5704-R, MAY 2008 — Previous editions are obsolete. — APD PE v1.00

*Figure B-2. Example completed DA Form 5704-R.

APPENDIX C
TRAINING SCHEDULES

To aid in the individual training phase, training schedules for the courses in pistol marksmanship training are described in this appendix. These schedules are based on the desirable number of training hours for a pistol course. They should be used as a guide in preparing lesson plans; conditions may require a longer or shorter period to complete the training. When time is available, additional training should be included in the schedule. When suggested equipment and training aids are not available, the best that are available should be improvised or substituted. Each firer should be allowed 50 rounds for instructional firing and 40 rounds for record firing.

C-1. 9-MM SEMIAUTOMATIC PISTOL, PRACTICE OR INSTRUCTIONAL FIRING COURSE (12 HOURS)

Period	Hours		Lesson	References	Training Facilities	Training Aids
	Peace	Mobilization				
MECHANICAL TRAINING (4 Hours)						
1	2	2	Characteristics, disassembly and assembly, functioning, and care and cleaning.	TM 9-1005-317-10, TM 9-1005-325-10.	Classroom or field.	For each instructor: chalkboard, working model, projector and screen. For each man: cleaning equipment. For each group: table or suitable ground cloth.
2	2	2	Malfunctions, stoppages, immediate action, loading, unloading ammunition, and safety precautions.	TM 9-1005-317-10, TM 9-1005-325-10.	....do....	Same as period 1 plus ammunition display.
PREPARATORY MARKSMANSHIP TRAINING (6 Hours)						
3	3	3	Coaching, aiming, grip, positions, trigger squeeze (to include double-action), target engagement, and slow-fire exercise.	Chapter 2 of this manual.	....do....	For each man: one pistol with magazine. For all: E-silhouette.
RANGE FIRING (2 Hours)						
4	2	2	Instructional Firing Tables I through V, Combat Pistol Qualification Course.	Appendix A of this manual.	Live-fire range.	Equipment used in period 6 of the qualification course.

C-2. 9-MM SEMIAUTOMATIC PISTOL, QUALIFICATION COURSE (12 HOURS)

Period	Hours		Lesson	References	Training Facilities	Training Aids
	Peace	Mobilization				
MECHANICAL TRAINING (4 Hours)						
1	2	2	Characteristics, disassembly and assembly, functioning, and care and cleaning.	TM 9-1005-317-10, TM 9-1005-325-10.	Classroom or field.	For each instructor: chalkboard, working model, projector and screen. For each man: cleaning equipment. For each group: table or suitable ground cloth.
2	2	2	Malfunctions, stoppages, immediate action, loading, unloading, ammunition, and safety precautions.	TM 9-1005-317-10, TM 9-1005-325-10.	....do....	Same as period 1 plus ammunition display.
PREPARATORY MARKSMANSHIP TRAINING (4 Hours)						
3	2	2	Coaching, aiming, grip, positions, trigger squeeze (to include double-action), target engagement, and slow-fire exercise.	Chapter 2 of this manual.	....do....	For each man: one pistol with magazine, For all: E-silhouette.
4	2	2	Review and examination.	All previous references.	....do....	For all: all equipment used in previous periods.
RANGE FIRING (4 Hours)						
5	2	2	Instructional firing combat pistol qualification course, for practice with a coach or instructor.	Existing range regulations. Appendix A of this manual.	Pistol range.	For all: all equipment used for periods 3 and 4 plus scorecard and ammunition.
6	2	2	Record firing, Tables I through V, combat pistol qualification course.	Appendix A of this manual.	....do....	............do............

Glossary

APQC	alternate pistol qualification course
AR	Army regulation
CBRN	chemical, biological, radiological, or nuclear
CLP	cleaner, lubricant, preservative
CPQC	combat pistol qualification course
CTA	common table of allowances
DA	Department of the Army
EENT	end evening nautical twilight
EMNT	end morning nautical twilight
FM	field manual
HQ	headquarters
LSA	lubricating [oil], semifluid, automatic [weapons]
mm	millimeter
MOPP	mission-oriented protective posture
NATO	North Atlantic Treaty Organization
NCOIC	noncommissioned officer in charge
NG	[Army] National Guard
OIC	officer in charge
QTTD	quickfire target-training device
RBC	rifle bore cleaner
RH	right hand
TM	technical manual
TRADOC	[US Army] Training and Doctrine Command
USAR	US Army Reserve

REFERENCES

DOCUMENTS NEEDED

These documents must be available to the intended users of this publication.

AR 385-63	Policies and Procedures for Firing Ammunition for Training, Target Practice, and Combat. 15 October 1983.
DA Form 88	Combat Pistol Qualification Course Scorecard.
DA Form 5704-R	Alternate Pistol Qualification Course.
TM 9-1005-317-10	Operator's Manual for Pistol, Semiautomatic, 9-mm, M9 (1005-01-118-2640). 31 July 1985.
TM 9-1005-325-10	Operator's Manual for Pistol, Compact, 9-mm, M11 (1005-01-336-8265) and Pistol, Compact, 9-mm, M11 with Tritium Sights (1005-01-340-0096). 16 December 1993.
TM 9-1300-200	Ammunition, General. 3 October 1969.

SOURCES USED

These are the sources quoted or paraphrased in this publication.

AR 140-1	Mission, Organization, and Training. 1 September 1994.
CTA 8-100	Army Medical Department Expendable/Durable Items. 31 August 1994.
CTA 50-970	Expendable/Durable Items (Except: Medical, Class V, Repair Parts and Heraldic Items). 21 September 1990.
DA Pam 350-38	Standards in Weapons Training. 3 July 1997.
DA Pam 738-750	Functional Users' Manual for the Army Maintenance Management System (TAMMS). 1 August 1994.
FM 3-4	NBC Protection. 29 May 1992, with Changes 1-2, 28 October 1992-21 February 1996.
FM 3-5	NBC Decontamination. 28 July 2000.
FM 19-10	The Military Police Law and Order Operations. 30 September 1987.

FM 21-11	First Aid for Soldiers. 27 October 1988, with Changes 1-2, 28 August 1989-4 December 1991.
FM 21-75	Combat Skills of the Soldier. 3 August 1984.
TC 25-8	Training Ranges. 25 February 1992.
TM 9-1005-317-23&P	Unit and Intermediate Direct Support Maintenance Manual (Including Repair Parts and Special Tools List) for Pistol, Semiautomatic, 9-mm, M9 (1005-01-118-2640). 16 October 1987.
TM 9-1005-325-23&P	Unit and Intermediate Direct Support Maintenance Manual (Including Repair Parts and Special Tools List) for Pistol, Compact, 9-mm, M11 (1005-01-336-8265) and Pistol, Compact, 9-mm M11 with Tritium Sights (1005-01-340-0096). 1 March 1993.
TM 9-6920-210-14&P	Operator and General Support Maintenance Manual (Including Basic Issue Items List and Repair Parts List) for Small Arms Targets and Target Material. 2 June 1992.
TM 43-0001-27	Army Ammunition Data Sheets for Small Caliber Ammunition. 29 April 1994.

INTERNET SOURCES USED

The following Internet sources were used in the preparation of this manual.

U. S. Army Publishing Agency, http://www.usapa.army.mil

Army Doctrine and Training Digital Library, http://www.adtdl.army.mil

INDEX

ALTERNATE PISTOL QUALIFICATION COURSE SCORECARD

For use of this form, see FM 3-23.35; the proponent agency is TRADOC.

1. ID CODE[1]	2. UNIT	3. LANE	4. ORDER	5. GROUP	6. DATE (YYYYMMDD)
20080426LUS01	C CO 2/29 IN	3	4	5	20080426

7. TABLE I — DAY STANDING[2] — 1 Mag : 7 Rds

TIME	TGT	HITS
21 Sec	1	X
	2	X
	3	X
	4	M
	5	X
	6	X
	7	M
Table I Total Hits		5

8. TABLE II — DAY KNEELING[2] — Mag 1 : 6 Rds[3] — Mag 2 : 7 Rds

TIME	TGT	HITS
45 Sec	1	X
	2	X
	3	X
	4	M
	5	X
	6	X
	7	X
	8	X
	9	X
	10	X
	11	X
	12	X
	13	X
Table II Total Hits		12

9. TABLE III — DAY CROUCHING[2] — Mag 1 : 5 Rds[3] — Mag 2 : 5 Rds

TIME	TGT	HITS
35 Sec	1	X
	2	X
	3	X
	4	X
	5	X
	6	X
	7	X
	8	M
	9	X
	10	X
Table III Total Hits		9

10. TABLE IV — DAY PRONE UNSUPPORTED[2] — Mag 1 : 5 Rds[3] — Mag 2 : 5 Rds

TIME	TGT	HITS
35 Sec	1	X
	2	X
	3	X
	4	X
	5	X
	6	X
	7	X
	8	X
	9	X
	10	X
Table IV Total Hits		10

11. TABLE V — DAY CBRN CROUCHING[2] — 1 Mag : 7 Rds

TIME	TGT	GO	NO-GO
21 Sec	1	X	
	2		M
	3	X	
	4	X	
	5	X	
	6	X	
	7	X	
Table V Total Hits		6	1
GO or NO-GO[4]		✓	

12. TABLE VI — NIGHT CROUCHING[2] — 1 Mag : 5 Rds

TIME	TGT	GO	NO-GO
21 Sec	1		M
	2	X	
	3	X	
	4	X	
	5	X	
Table VI Total Hits		4	1
GO or NO-GO[4]		✓	

13. NOTES

1 Do not use personal information.

2 Firer receives the number of rounds required to fire a specific table. The OIC of firing sets procedures for loading and unloading.

3 Firer must change magazine rapidly.

4 Check GO or NO-GO box for whole table.
- Four hits are required for a GO on Table V.
- Two hits are required for a GO on Table VI.

14. RATING CALCULATOR

Enter total hits from Tables I through IV and add them to obtain a grand total:

TABLE I	5	
TABLE II	12	
TABLE III	9	
TABLE IV	10	+
GRAND TOTAL	36	=

15. RATING SCALE

GRAND TOTAL (BLOCK 14)	RATING	
36 - 40	EXPERT	X
29 - 35	SHARPSHOOTER	
24 - 28	MARKSMAN	
0 - 23	UNQUALIFIED	

16. GRADER'S INITIALS	17. DATE (YYYYMMDD)	18. OIC'S INITIALS	19. DATE (YYYYMMDD)
KGF	20080426	CAH	20080426

DA FORM 5704-R, MAY 2008 — Previous editions are obsolete. — APD V1.00

*Figure B-2. Example completed DA Form 5704-R.

ALTERNATE PISTOL QUALIFICATION COURSE SCORECARD

For use of this form, see FM 3-23.35; the proponent agency is TRADOC.

1. ID CODE [1]	2. UNIT	3. LANE	4. ORDER	5. GROUP	6. DATE *(YYYYMMDD)*

7. TABLE I
DAY STANDING [2]
1 Mag : 7 Rds

TIME	TGT	HITS
21 Sec	1	
	2	
	3	
	4	
	5	
	6	
	7	
Table I Total Hits		

8. TABLE II
DAY KNEELING [2]
Mag 1 : 6 Rds [3]
Mag 2 : 7 Rds

TIME	TGT	HITS
45 Sec	1	
	2	
	3	
	4	
	5	
	6	
	7	
	8	
	9	
	10	
	11	
	12	
	13	
Table II Total Hits		

9. TABLE III
DAY CROUCHING [2]
Mag 1 : 5 Rds [3]
Mag 2 : 5 Rds

TIME	TGT	HITS
35 Sec	1	
	2	
	3	
	4	
	5	
	6	
	7	
	8	
	9	
	10	
Table III Total Hits		

10. TABLE IV
DAY PRONE UNSUPPORTED [2]
Mag 1 : 5 Rds [3]
Mag 2 : 5 Rds

TIME	TGT	HITS
35 Sec	1	
	2	
	3	
	4	
	5	
	6	
	7	
	8	
	9	
	10	
Table IV Total Hits		

11. TABLE V
DAY CBRN CROUCHING [2]
1 Mag : 7 Rds

TIME	TGT	GO	NO-GO
21 Sec	1		
	2		
	3		
	4		
	5		
	6		
	7	☐	☐
Table V Total Hits			
GO or NO-GO [4]		☐	☐

12. TABLE VI
NIGHT CROUCHING [2]
1 Mag : 5 Rds

TIME	TGT	GO	NO-GO
21 Sec	1		
	2		
	3		
	4		
	5		
Table VI Total Hits			
GO or NO-GO [4]		☐	☐

13. NOTES

1 Do not use personal information.

2 Firer receives the number of rounds required to fire a specific table. The OIC of firing sets procedures for loading and unloading.

3 Firer must change magazine rapidly.

4 Check GO or NO-GO box for whole table.
. Four hits are required for a GO on Table V.
. Two hits are required for a GO on Table VI.

14. RATING CALCULATOR

Enter total hits from Tables I through IV and add them to obtain a grand total:

TABLE I		
TABLE II		
TABLE III		
TABLE IV		+
GRAND TOTAL		=

15. RATING SCALE

GRAND TOTAL *(BLOCK 14)*	RATING	
36 - 40	EXPERT	☐
29 - 35	SHARPSHOOTER	☐
24 - 28	MARKSMAN	☐
0 - 23	UNQUALIFIED	☐

16. GRADER'S INITIALS	17. DATE *(YYYYMMDD)*	18. OIC'S INITIALS	19. DATE *(YYYYMMDD)*

Made in the USA
Monee, IL
04 November 2026

67383944R00039